Ethical Prospects

Ethical Prospects

Economy, Society and Environment

Volume 1

Editor in Chief
Laszlo Zsolnai

Associate Editors
Zsolt Boda
Laszlo Fekete

Springer

Editor in Chief

Laszlo Zsolnai
Business Ethics Center
Corvinus University of Budapest
Budapest, Hungary
E-mail: laszlo_zsolnai@interware.hu

Associate Editors

Zsolt Boda
Business Ethics Center
Corvinus University of Budapest
Budapest, Hungary
E-mail: bodazsolt@yahoo.com

Laszlo Fekete
Business Ethics Center
Corvinus University of Budapest
Budapest, Hungary
E-mail: lfekete@ella.hu

ISBN 978-90-481-8206-0 e-ISBN 978-1-4020-9821-5

Printed on acid-free paper

9 8 7 6 5 4 3 2 1

springer.com

Ethical Prospects

Ethical Prospects aims to present and summarize new perspectives and leading-edge results in ethics reflecting on interconnected economic, social and environmental issues. It reports on innovative practices and policy reforms and provides a forum for discussion about ground-breaking theories. The main function of the yearbook is to present ideas and initiatives that lead toward responsible business practices, policies for the common good and ecological sustainability. It seeks to form a value-community of scholars, practitioners and policymakers engaged in genuine ethics in business, environmental management, and public policy.

International Advisory Board

All correspondence should be sent to

Laszlo Zsolnai
Professor and Director
Business Ethics Center
Corvinus University of Budapest

Budapest 5, P.O. Box 489
1828 Hungary

Email: laszlo_zsolnai@interware.hu

Contents

Laszlo Zsolnai
Editor in Chief

Introduction: Ethical Prospects

It is my pleasure to introduce the "Ethical Prospects: Economy, Society, and Environment" volume 1. It is a product of 27 scholars and practitioners from Europe, America and Asia. Contributors represent a diversity of fields including business ethics, philosophy, organizational science, systems theory, finance, management, economics, political science, and ecology.

In his paper "The Good Company," American business ethicist *Edwin M. Epstein* of the University of California at Berkeley investigates the possibility of the "good company" and the mission of business ethics in the contemporary globalised economy. He points out that notwithstanding the best of intentions, there are *constraints* on the ability of companies and their managers to *achieve sociably responsible behaviours*.

Epstein argues that we must rely on legal processes, both nationally and trans-nationally to establish the "rules of game" for desired corporate behaviours. In addition, we must encourage business organizations to engage in meaningful self-regulation of their activities on an industry or regional basis and thereby obviate the necessity for additional public policy initiatives. Epstein believes that ethical norms and aspects of CSR embodied within the concept of the corporate social policy process will assist managers in assessing the broader societal implications of their decisions. Vigilant media monitoring and vigorous citizen oversight of corporate action are essential to assure that business firms are both rendered accountable for the consequences of their actions and exposed to public expectations of "best practices" of business performance.

Epstein concludes that a more realistic goal to stimulate the spread of "good companies" which connotes the achievement of a certain type of corporate perfection is to strive for "better companies," organizations which continuously seek to perform the economic functions for which society relies upon them in a manner which optimizes the firm's utility to the diverse stakeholders affected by their actions and minimizes the deleterious effects (what economist term "externalities") of their operations.

In his paper "Dialogic Ethics for Business" Hungarian philosopher *Imre Ungvari Zrinyi* of "Babes Bolyai" University of Cluj calls for adopting a dialogic perspective in business in general and in business ethics in particular. Referring to the work of Martin Buber, Emmanuel Levinas, Hans-Georg Gadamer, Karl-Otto Apel and Jürgen Habermas he urges to surpassing the monological standpoints in business activities.

To conceive business as a system of communication and negotiation through which different interests could enhance societal well-being as well as meeting the personal needs of the negotiating parties places the dialogic attitude in the center of business activity – argues Ungvari Zrinyi. The main concerns for taking dialogue as a model for honest human relations are the followings: (i) contemporary societies are subject to differentiation and interdependence so major purposes can no longer be substantiated monologically, they have become subjects of a dialogue; (ii) people lead their life in a permanent interaction between 'self' and 'other', they must choose between the conflict and the dialogue of intentions and differences; (iii) an essential trait of sociality – as the German philosopher Max Scheler stressed –, is exactly to be present 'for one another'; (iv) dialogism provides us with a 'situated ethics' that represents an alternative to a coercive moral absolutism; (v) dialogic conceptions focus on the relational and discursive origin of ethics rather than communal, individual or rational one; and (vi) by including dialogical conceptions in the ethical discussion about business activity we hope to reach an insightful, more creative, responsive and responsible praxis of management.

In their paper "Management Education – A Path to Business Integrity?" Irish management scholar *Eleanor O'Higgins* of the University College Dublin and London School of Economics and French organizational scientist *Yvon Pesqueux* of CNAM Paris examines how contemporary management philosophies give rise to the practices that have created scandals and business failures. They depict the current role of management education in perpetuating the situation. Also they ask whether it is possible for management education to reverse its role to play a part in creating a more honourable business environment.

O'Higgins and Pesqueux argues that "managerialism" has created a void where ethical contents should be, and management education has perpetuated this situation, either actively by mimicking the managerialist model, or passively, by retreating into pedantic science. The situation will be difficult to reverse, but a transformation in the underpinning and practices of management education will enable it to make a contribution to business integrity. Suggestions by O'Higgins and Pesqueux center on new links between the business school and society, between research and practice and between the business school and business itself. They call for changes in internal processes and practices in business schools that will enable the new links to flourish.

In his paper "Value Creation as the Foundation of Economics" American philosopher *Robert Elliott Allinson* of Soka University of America argues that self-interest economics is fundamentally flawed and needs to be replaced by a value based economics. One thread of his argument is an attempt to show why the fundamental philosophical notions of Adam Smith, taken as an illustration of self-interest economics, cannot lead to an equitable society. According to Allinson the real problem for Smith is that one must treat economic relations as being between non-persons in order to submerge ethical impulses under economic ones. In so doing, not only does one "ex hypothesi" treat others unethically, it also, according to Aristotelian and Confucian notions, makes oneself unethical and helps to co-create an unethical society.

The second thread of Allinson's argument is the theory of value based economics rather than profit based economics. The theory of value creation says that all values, aesthetic, social and ethical are to be part of one's motivation for action in life. Action is for value creation, not for profit. Profit is the side-effect, not the goal. We should build ethics and aesthetics into the business in the first place. Value based economics may be defined as the ownership or the use of capital investment, labour or land to produce a product or provide a service that creates a positive social value without creating a disvalue which is of greater harm than the value produced.

Allinson emphasizes that by following a self-interest and profit based economics, one inevitably shapes one's character, but in this case, one shapes an unethical character. This unethical character, moulded by the daily pursuit of profit, must come to battle with one's ethical impulses. Allinson's point is that a system founded upon self-interest is inherently unethical as a system, not by its abuse. It is unethical in its proper use.

Economics cannot exist for the sake of itself: it must exist for the sake of something else. Our works, our labors, our investments, all exist for the sake of a higher end. As Joan Robinson once asked, 'Here we come upon the greatest of all economic questions, but one which is never asked, what is growth for?' Allinson's answer is that proper growth serves the end of the creation of positive value and the reduction of excess value, disvalue or negative value.

In his paper "Buddhist Economics for Business," Hungarian business ethicist *Laszlo Zsolnai* of Corvinus University of Budapest explores Buddhist economics for transforming business toward a more ecological and human form. Buddhist economics is centered on want negation and purification of the human character. It challenges the basic principles of Western economics, (i) profit-maximization, (ii) cultivating desires, (iii) introducing markets, (iv) instrumental use of the world, and (v) self-interest based ethics. In Zsolnai's reconstruction Buddhist economics proposes alternative principles such as (I) minimize suffering, (II) simplifying desires, (III) non-violence, (IV) genuine care, and (V) generosity.

Zsolnai stresses that Buddhist economics is not a system. Rather it represents a strategy, which can be applied in any economic setting. It helps to create livelihood solutions which reduce suffering of all sentient beings by practicing want negation, non-violence, caring and generosity. Today's business model is based on and cultivates narrow self-centeredness. Buddhist economics point out that emphasizing individuality and promoting the greatest fulfillment of the desires of the individual conjointly lead to destruction.

Happiness research shows – warns Zsolnai – that not material wealth but the richness of personal relationships determines happiness. Not things but people make people happy. Western economics tries to provide people with happiness by supplying enormous quantity of things. But what people needs are caring relationships and generous love. Buddhist economics make these values possible by direct provision. Peace can be achieved by non-violent ways. Wanting less can substantially contribute to this task and make it happen easier. Permanence, that is, ecological sustainability requires a drastic cut back of the present level of consumption and production globally. This reduction should not be an inconvenient exercise of self-sacrifice. In the noble ethos of reducing suffering it can be a positive development path for business.

In their paper "The Role of Government in Corporate Social Responsibility" *Laura Albareda* and *Josep Maria Lozano* of ESADE Business School Barcelona, *Antonio Tencati* and *Francesco Perrini* of Bocconi University Milan, and *Atle Midttun* of Norwegian School of Management, Oslo contribute to understanding of the government's role promoting Corporate Social Responsibility (CSR) as an answer to the debate that has arisen within the political and global context during the last decade. Their paper: (1) sets out the research findings and the various factors governments or the academic literature consider to be key drivers behind public initiatives for fostering CSR; (2) proposes a relational framework as a model for analysing the various approaches taken by governments and looks at the various public policies that governments may adopt to foster CSR. These policies are classified by the agents involved, and in particular by their relationships; (3) explores the relationship among social, economic and environmental context and the development of CSR governmental approaches; and (4) sets out the governments', companies', and NGOs' perceptions as to what role governments ought to play in fostering CSR.

The analysis regarding public CSR policies in three European countries (Italy, Norway, and the United Kingdom) indicates some key issues as follows. (i) At the international level, there is a common discourse on CSR, shared by various international organizations such as the European Union, the UN Global Compact, or the International Labour Organization and the international CSR network (CSR Europe, World Business Council for Sustainable Development, Business for Social Responsibility, and Forum Empresa among others). However, the application of this discourse into specific governmental policies and programs varies across countries. Governmental policy frameworks and policy implementation vary regarding their vision, objectives and priorities; their scope (international versus domestic); organizational structures; initiatives; policy mechanisms and programs; levels of implementation at regional and local levels. (ii) CSR policies are directly linked to the globalised context in which governments find themselves; the increasing power of companies over governments; and the crisis in the Welfare State. In this respect, the authors consider that the links need to be analyzed between public CSR policies and social, economic and environmental indicators of economic growth, unemployment, inequality in income distribution, population ageing, competitiveness/innovation and environment in each country. (iii) In many countries, CSR public policies focus mainly on relations between governments and companies, although in some cases they cover policies addressing the relationships between government, companies, and civil society. If CSR is defined by its multi-stakeholder and partnership characteristics, this relational aspect needs to be a basic aim of governmental policies and programs. The complex relationships between social agents reveal a new role for governments as promoters and catalysts for socially, environmentally responsible policies in partnership with social actors.

The paper "Sustainable Systems Implementation" by American investment analyst *Frank Dixon* (Innovest, New York) concerns improving unsustainable business practices. Dixon's sustainability approach combines traditional corporate responsibility efforts with system change efforts. Mid-level system change is focused on specific sectors, stakeholder groups, or environmental or social issues. High-level system change is focused on improving overarching economic, political and social systems.

Dixon introduces principles which show important activities and viewpoints needed to evolve human systems into sustainable forms. These principles are as follows: "Systems Thinking", "Simplicity", "Inevitability", "Evolution not Revolution", "Be Willing to Question Everything", "Abide by Nature's Laws", "Non-Judgment", "Don't Expect Individuals to Change", "Carrot not Stick", "Responsibility", "Focus on Results", and "Visionary, Courageous Leadership." He warns that pressure from the financial community to supply ever-increasing financial returns obscures the fact that the primary obligation of firms is to society, not to shareholders. Companies do not have an inherent right to exist. That right must be earned. Providing valuable products and services cannot be used as a justification for ongoing negative impacts, as it often is. Even if regulations allow firms to negatively impact society, as they often do, firms have an obligation to go beyond this and mitigate all impacts.

Business leaders can take a seat at the system change table – suggests Dixon. Rather than having system changes for which they are not prepared wash over them, leaders can help to shape practical, reasonable changes that improve business and society. Even on a purely financial basis, this makes sense. As environmental and social issues become more financially relevant, the fiduciary obligation to maximize shareholder returns demands that firms mitigate negative impacts to the greatest extent possible.

In his paper "Public versus Private Domain: Knowledge and Information in the Global Communications Network" *Laszlo Fekete* of the Business Ethics Center at the Corvinus University of Budapest argues that the prevailing intellectual property laws, international treaties and agreements are captured by the vested interest of corporate rights holders in implementing strict and comprehensive proprietary rights over knowledge and information goods.

The proliferation of private ordering by means of licenses and standard form contracts brings about the pre-emption of the provisions of the prevailing intellectual property laws. The prevalence of contractual arrangements in the global communications network would be welcome if it were relied upon the voluntary, rational, and deliberate consent of parties. Fekete observes that the great majority of contracts completed in the online world does not depend upon the meeting of the minds of the parties. Standard form contracts, license terms and conditions are deliberately drafted on a take-it or leave-it basis. Though, this general business practice is not devoid of paradox: private ordering regime offers an appealing perspective to wither away the public domain in order to maximize the economic benefits of corporate rights holders as well as to pre-empt the biased and restrictive provisions of the recent intellectual property laws by means of providing free – or at least less constrained – access to knowledge and information goods.

Fekete concludes that as the production, use, and dissemination of knowledge and information are essentially cooperative enterprise, it is quite reserved proposal to appeal to rights holders for a gesture of goodwill, virtue, and decency without scrutinizing the legitimacy, social fairness, and economic efficiency of the current proprietary regime over knowledge and information goods as defined by the law. Well-defined and exclusive property rights are indispensable for promoting the allocative efficiency of the market. Though, the deference to this well-established paradigm about the primacy of private goods over public goods is neither self-evident nor particularly productive in case of knowledge and information.

In his paper "Changing the Scoring System for the Game of Economic Life" British economist *James Robertson* suggests that people concerned about the ethics of business should see the money system as the scoring system for the game of economic life and should consider how they could help to change it to get a better game. He stresses that the state should carry out its main operational monetary and financial responsibilities in ways that will distribute the value of common resources among all citizens and reduce or even abolish taxes on earnings and profits from providing useful goods and services. This can create a new framework of prices which reward the market economy for delivering outcomes which combine economic efficiency with social justice and environmental care. People in business will then find not only that their businesses need to be less closely regulated by the state but that they themselves are troubled less by conflicts between the ethical values held by themselves, their families and their friends and the money values that must necessarily shape their business behaviour if they are to survive and succeed.

The changes in the money system outlined in Robertson's paper could help to create a new direction of economic development. Attention will shift to creating well-being for people and the Earth; to enabling people to develop their capability, rather than reinforcing their dependency; and to conserving the Earth, rather than transforming its resources as rapidly as possible into money. The fairer sharing of the value of common resources will help to decentralise power and wealth – both by giving a fairer deal to people in their own places and by requiring rich and powerful people and corporations and nations to bear their full share of the environmental and social costs of centralisation. The new framework of monetary and financial incentives will automatically harness self-interest to common interest within and between nations. Reducing the present mismatch between money values and ethical values will reduce the need to swim against the financial tide in order to act in accordance with their ethical values.

In his paper "Ethical Banking" Belgian banker *Frans de Clerck* of the Triodos Bank Group introduces working models of "ethical, sustainable, social, alternative, development or solidarity" banking and finance. They share a values-driven approach to their business, at odds with the conventional commercial banking and finance industry. Although private community and development banks, microfinance banks, ethical, environmental and social banks and ethical funds differ in terms of focus, accents, clients, products and business culture, they have in common to practice banking and investment with a human development mission.

Some questions require continuous attention – warns de Clerk. Will this emerging financial business sector be able to achieve the relative scale and the professionalism to challenge the dominance of mainstream finance? Will the exceptions of the financial industry become the exceptional and a factor in modern society? Will a profound way of dealing with ethical choices be overruled by the superficiality of business development – also in ethical banking? Can ethical banking as a process with an instrumental character avoid becoming institutionalized? Can ethical banking be a portal for trust forces, morality and responsibility to feed money processes and the financial system with basic values and practices that can counteract uncontrollable growth? De Clerk believes that ethical banking could become a significant factor in society, less in terms of volume and more through the human added value it delivers.

The "Opinions" section focuses on the *problem of future generations*. Our unsustainable economy, the depletion of natural resources, the destruction of eco-systems and the long-range effects of modern technologies, such as nuclear energy or genetic modification, obviously affect the quality of life of future generations. However, the rights and interests of yet unborn people are not represented in the decision making processes and institutional setting of today's society.

Associate editor, *Zsolt Boda* asked a distinguished panel of academics and practitioners to answer the following questions: What kind of reforms could be possibly proposed so that the regulatory institutions of the global economy represent the rights and interests of the coming generations? What can business do in order to consider its effects on future generations? Are there principles, norms and decision making rules to follow? Or is the respect for the future is a challenge still to be met by business ethics thinking?

In his response, "Representing Future Generations" *Jakob von Uexkull*, Founder of the World Future Council states that the interests of future generations need institutional representation, either by an organization with moral power such as the Councils of Seers Into The Future in pre-colonial India or the World Future Council today. He states that the business of business should be service to society, not preserving privileges and maximizing profits. Clearly, profitability is necessary to cover costs and living expenses but that is very different from maximizing profits by externalizing costs at the expense of our children. Businesses active in areas which entail substantial future risks should be legally changed from limited liability corporations to foundations mandated to act for the common good.

In market economies, markets are servants. Today we have market societies where markets masters – with terrible consequences. But we can decide to follow the rules of ethics rather than the rules of profit – suggests Uexkull.

In his response, "A Speechless Mass Behind Sustainability" *Benedek Javor*, representative of a Hungarian civil society organization "Védegylet" in Budapest emphasizes that as human and social rights are a widely accepted and secure basis for incorporating social costs into the economic activity, rights of future generations is the final argument behind the claim for a sustainable system. Unborn people should be let in under the protecting umbrella of basic human rights.

Javor recalls the case of his own organization, Vedegylet (Protect the Future!). In 2000, they initiated to set up the office of the Ombudsman for Future Generations in Hungary. The bill prepared by Laszlo Solyom, the present president of Hungary, and member of the board of Védegylet at that time, proposed the establishment of an institution which sets the reference base of its interventions in the effective Hungarian legal system so that the commissioner's recommendations and procedures be built on a "hard" legal base. Hard debates ended in November 2007 an all-party introduction of the bill on the Ombudsman for Future Generations. Soon, the Hungarian Parliament accepted the bill, creating an internationally unique, model-like institution for the protection of the interests of future generations. And in May 2008 the new ombudsman was finally elected. His appearance in the political decision-making processes really makes a difference, and helps Hungary to ensure a more livable natural and human environment for future generations.

In her response, "The Right of Future Generations" Finnish civil servant, *Paula Tiihonen* of the Committee for the Future of the Finnish Parliament presents the case of Finland in protecting the right of future generations. In 1993 the Finish Parliament decided to establish a special committee to deliberate problems of our common future, and thus also the rights and responsibilities of future generations. It was tentatively named the "Committee of the Future", but in retrospect I would say it got off to a very good start in that at its very first meeting the 17 members (all parliamentarians, representing the full spectrum of political parties) decided to change the name to the "Committee for the Future". The change of preposition demonstrated that they wanted their committee to be active rather than passive; to be for the future not against it, to take an innovative attitude to society, science and technology.

Having a mission and working hard during the last 13 years, the committee has taken its place in the Finnish parliamentary system as an innovative political body. It has created a new forum that works at the core of the parliamentary system and has demonstrated that parliamentary measures can still be used to take the initiative within democracy.

In her response "The Legacy of Business Leaders" former Harvard Business School researcher *Laura Nash* states that the extension of our moral vision to the right of future generations is one of the exercises that characterizes the uniquely human capacity to transcend the boundaries of time, knowledge and physicality.

She observes that the business leader that can articulate and execute a strategy of "just enough" on environmental responsibilities will create a legacy most conducive to securing the rights of future generations. Getting there depends on the cultivation of more energizing assumptions that sheer cutbacks, and a special kind of emotional intelligence about future generations. Secondly, as leaders and governing boards deliberate an ethic of natural resources sustainability, their framing question should move beyond conservation to integrate a full articulation of conception of the good life, concept that would include generativity.

Nash labels "Legacy" the drive of business leaders who sustain multiple forms of success in their lives. It is a good model for understanding our legacy for ensuring rights to future generations. It can be observed a surprising and satisfying emotional payback in the way some leaders approach Legacy. For them, Legacy concerns motivate the creation of a platform that helps others succeed in the future.

Nash warns that knowing "just enough" is needed at a deeply ethical and emotional level. Religious and secular exercises in mindfulness help cultivate such felt satisfaction, and with it the patience and maturity to think through the complex technological, economic, and social challenges that environmental sustainability poses. The businesses that lead in cultivating such emotions and perspective will create a legacy worth banking on. Using these suggested perspectives, businesses can not only face the idea that they have an obligation to factor in the rights of future generations, they can create a viable approach to doing so.

In his response "Enhancing the Quality of our Decisions for Nurturing a Sustainable World" Indian scholar *J. M. Sampath* of Arpitha Associates in Bangalore sets a framework for decision making respecting future generations. He says that every business decision should be re-traceable to the vision of the business itself. Those in positions of influence should be able to see that every action of theirs contributes to achieving the ultimate vision, taking into consideration various elements that connect the decision to the vision. When a decision does not yield the desired results, it is almost certain that one or more elements would have been missed out or not considered enough. Thus it becomes important for leaders to be mindful of all these elements during decision-making. The absence of this alignment results in conflicts that could affect the business and larger system in several ways.

Sampath emphasizes that over the years we have taken the values, process, clarity and standard for granted and all our goals, strategies and choices have been focused on short term benefits. We have manipulated every context while making decisions only from the perspectives of short term gain and not from the perspective of long term sustainable growth. It is time that leaders, while making decisions, consider all the elements without having to compromise any and this would in the long run, help us reverse the degeneration that is already set in today's businesses. Each of us need to understand that vision without values can be risky, values without vision will go nowhere and vision with values will result in the evolution of a world which will be more sustainable and easier to live in.

The "Debate" section is about the paper "Republican Liberalism versus Market Liberalism" by Swiss philosopher and business ethicist *Peter Ulrich* of the University of St. Gallen. He highlights the fundamental difference between republican liberalism and market liberalism. His major points are summed up as follows. First, as motivational base, an appropriate *civic virtue* with the consequence of a republican business ethos. Second, as the constitutive order principle, fully established *civic rights* including the socio-economic preconditions of a really free and decent living. Third, as an economic-ethical consequence, a *civilized market economy.* Ulrich stresses that republican liberalism does not contradict an efficient market economy but has to be conceived as a societal precondition of legitimate and life-serving business activities.

In his comments *Stephen B. Young*, Global Executive Director of The Caux Round Table in New York argues that Ulrich makes a category mistake when he conflates the business enterprise with those who have responsibility for a public order. Business is not government; it should not be run on principles of quid-pro-quo, muddling-through, or grid-lock until a consensus emerges among the citizenry as to what to do. The point of a business is to take entrepreneurial risks and doing so mandates making decisions and cutting opportunity costs. In responding to the desires of some, business must reject the desires of others. Government has its own office to perform, separate from that assigned to business. Government needs its own set of ethical principles to keep its use of power just and responsible. When we are assessing the advantages and disadvantages of our versions of civilization, we should not forget the golden eggs when we critique the goose. Capitalism, free markets, business enterprise, has given us health, wealth and power that our ancestors never imagined – adds Young.

In his reflection "Who Can Civilize the Market?" French economist *Jean-Pierre Galavielle* of the Université Paris 1 (Sorbonne) focuses on the new phenomenon of our age: *ungovernability*, which means that public authorities are no longer capable of bearing their responsibilities. Galavielle believes that the crisis of governability appears due to a loss of legitimacy of the socio-political system. He argues that the problem of the imbrications of the market and democracy within contemporary capitalism will not be resolved as long as there will remain the myth of the dictatorship of the "government of things". Civilizing the market means that the politician must urgently regain his place as a democratic regulator of production, exchange and distribution of wealth.

In his response "Is Ethics Integral?" Hungarian philosopher, *Alpar Losoncz* of the University of Novy Sad argues that a shift from the "possessive" to the "republican" liberalism requires a stronger sense of community than now prevails. Republicanism articulates the continuity between the private and public ethics; in fact, it questions the dividing of the freedom, and requires a strong sense of common good. The indivisibility of the freedom is in line with the claiming of the effective and participative citizenship.

The argumentation of the business ethics – warns Losoncz – could not be directly translated into the political field. The business domain is only one order among others in a given social formation. Thus, it is just part of society and only indirectly is charged with responsibility for maintaining the cohesion of the wider society. In the society whose common interest is constituted, reproduced and transformed in and through heterogeneous processes we should take into account the autonomy of different spheres such as politics and economy. For this reason, he is skeptical in relation to any concept of "corporative citizenship". After all, we are forced to permanently explore and evaluate the changing relationships, moreover, the non-correspondences amongst the economic, non-economic forces, and ethical considerations.

In his reflection "Which Role for Business Ethics?," ethicist *Gerhold K. Becker* of Assumption University Bangkok and Hong Kong Baptist University underlines that Ulrich seems to present his conception of an integrated economic ethics as the super-theory that can absorb all rivals and transform itself into a foundational theory of the good society. But a case in point is the underdetermined relationship between the various types of rights (basic rights, civic rights, rights people are supposed to have as economic citizens, and human rights), their moral foundation, their hierarchical ranking within Ulrich's conception of the good life, as well as their institutional back-up in positive law.

Becker stresses that in the age of globalization, the ever growing circle of the affected parties is no longer restricted to nation states, their citizens, and domestic policies, but includes all people, regardless of social status, communicative ability or economic power; – it may even transcend humanity and include "nature" (ecological systems, higher animals). Conceptions of the well-ordered society inspired ethical thought throughout human history and were the driving forces in the process of civilization without ever achieving consensus among all stakeholders. For this reason Becker suggests a more modest attitude towards the capability of business ethics, it may nevertheless facilitate a more accurate picture of versions of applied ethics. While they deliberately refrain from projecting large vistas of comprehensive conceptions of the well-ordered society, they are certainly rooted in strong moral convictions that enable them to identify wrong developments and propose feasible alternatives.

In his reply, "Republican Liberalism and its Implications for Business Ethics" *Peter Ulrich* restates that republican liberalism is the proposal of a balanced conception that includes both a strong but neutral political order enabling a "reasonable pluralism" of conceptions of the good, and responsible citizens with a *republican ethos*, which means free citizens who recognize their shared responsibility for the well-ordered *res publica* and do not cleave off their business life from their civic identity and integrity. The resulting idea is *economic citizenship*, including an individual moment of citizenship responsibilities and an institutional complement of strong citizenship rights.

Ulrich calls for demystifying the deeply rooted economism of our age. He believes that neither the defence of the false "morality of the market" nor an approach of business ethics as applied ethics is adequate. The core problem of the latter is that it is conceived as a tool of social engineering for ethically good ends under the "given" business conditions of the existing market economy without reflecting these conditions themselves to the end. Ethics within the frame of free market capitalism is no longer enough today – what is needed is a comprehensive ethics of the preconditions of a legitimate and life-serving market economy.

Ulrich hopes that enriching business ethics with the republican spirit of true liberalism might be a more promising idea than to wait and see until the precarious normative foundations of the whole system will have lost their last legitimating force. Markets cannot "know" for what and for whom they shall be efficient – we have to "tell" them.

It is our hope that the papers presented in the *Ethical Prospects* volume 1 will provide good food for thought and inspiration for action for those who are engaged in the never ending "business" of ethics.

PART 1

New Perspectives and Findings

Edwin M. Epstein
University of California, Berkeley

The Good Company

In the hour when an individual is brought before the heavenly court for judgment, the person asked: Did you conduct your business affairs honestly?

(Babylonian Talmud, *Shabbat* 31a)

Falsifying and cheating with the weights is forbidden.
While I am weighing the goods, God is measuring me.

Latin Inscription, Sponza Palace, Dubrovnik, Croatia

For millennia, religious teachings and secular ethics have sought to nurture socially beneficial economic behavior. The canonical literatures of virtually every religious tradition as well as Western and Eastern philosophy are replete with precepts and admonitions regarding what constitutes ethical behavior in the economic realm.

To be sure, these teachings until relatively recently, have focused on the behavior of individuals vis a vis one another. Concerns about organizational entities, which aggregate human energy and diverse resources to undertake economic activity – the production and distribution of goods and services – is at best a century and a half old. To sharpen the point more finely, it is the emergence of large-scale business organizations in the last third of the nineteenth century both within Europe and the United States that gave rise to concerns about "corporate social responsibility." Early on, it became apparent that these "companies," "trust," "enterprises," in addition to being highly effective structures for developing complex industrial economies, were very significant social, political, and cultural institutions – individually and collectively – possessing considerable power and profoundly affecting all aspects of society from the local to the global. How to harness the energies of these powerful corporate engines for socially efficacious purposes yet constrain and render them accountable for their at times deleterious impacts on other sectors of society and assure they would be "Good Companies" has been an objective of public policy and an enduring private concern on the part of intellectuals, religious authorities, and ordinary citizens since the emergence of the "mega-corporation" in the nineteenth century.

1. The Contemporary Setting

This objective/concern has become all the more important after World War II, and especially in the past two decades following the end of the Cold War. During the latter period, we have witnessed the economic emergence of Big (China, India, South Korea) and Little (all the rest) Tigers in Asia, the expansion and fitful development of the European Union, combustibility in the Middle East, political and economic transformations in Latin/South America, and the ongoing economic and social agony of sub-Saharan Africa. "Globalization," the "Energy Crisis" and "Technological Revolution," and "Social Transformation" have become the buzz-word descriptors and the reality of the era. Let us briefly consider each of the phenomena and relationship to companies, good and otherwise:

- *Globalization* – Transnational business activity is hardly a new phenomenon as to anyone who recalls the Silk Road and its two millennia of East-West commerce on its routes or the European trading companies which aggressively spanned the globe with their economic and political colonization endeavors from the fourteenth to the twentieth centuries as did their latter-day American counterparts in the nineteenth and twentieth centuries. What is new is the advent of a truly world-wide trading regime made possible by modern communications and transportation technologies which have created a "woven world of distant encounters and instant connections" (Yergin and Stanislaw, p. 369) from which no place, however, geographically remote, is removed. Its critics notwithstanding, globalization is here to stay and transnational enterprises whether private or public, are and will remain the catalysts and prime instrumentalities of this process.

- *The Energy Crisis* – The dependence on fossil fuels by industrialized countries has dramatically altered the geo-political equation since the early 1970s when OPEC emerged as a formidable force. Countries (some barely two generations old and the products of post World War II decolonization) long dominated by Western powers have asserted their independence and exercised control over their energy resources – in some cases by nationalization and in other instances by requiring the renegotiation of exploration, production, refining and transportation leases with the petroleum majors working within their borders. The role of these companies – economic and political – and their relationships to the regimes controlling these newly-emergent nations have become increasingly complex, inevitably raising questions concerning whose interests they were serving and whether, indeed, inherent conflicts of interest exist. Unprecedented levels of company profits at a time when consumers, at least in the United States, are experiencing equally unprecedented price levels have exacerbated these issues.
- *Technological Revolutions* – High-tech, biotech and simply technological developments in traditional economic sectors have radically altered our global political economy. Nuclear power, computer chips and genetically engineered agricultural and pharmaceutical products are manifestations of the technological revolutions which have affected virtually every human being for good and for ill. Corporations – frequently in relationship with governmental entities – are the

catalysts and propellants of this revolution which has dramatically improved the quality of life for many persons while diminishing that quality for other global residents.

- *Social Transformations* – The three above-mentioned phenomena contribute to and are part of the Social Transformation which has characterized the past half-century. The shedding of colonial domination, the creation of the United Nation as a forum for less-developed countries to influence international developments and consciousness and the impact of communications technologies ranging from computers to short-wave radios have contributed to an awareness of deprivations and "entitlements" on the part of persons throughout the world. The maldistribution of the benefits of our global political-economy roils politics at international, inter-state and intra-state levels, devolving down to local communities. Transnational companies are key players in this changing social dynamic as utilizers and "transfer agents" of the physical, human, financial, and intellectual resources of our global society. As such, through their policies, processes and practices, they play an important role in facilitating global peace or exacerbating global conflict.

2. The Modes of Social Control

Given these developments which frame the contemporary global political economy, how can civil society promote corporate behavior which is beneficial rather than harmful? Is the goal of the "Good Company" utopian or at some level achievable? And what is the "art of the possible" in this endeavor? There exist a variety of modes, better still, processes for framing social behavior at both individual and organizational levels. In descending order of formality, they are: law/legal systems; affinity group regulation; self-regulation; mores, traditions customs, at times embodied in ethical norms; media monitoring; and civil society activism. Let us look at each of these modes of social control.

- *Law*, for these purposes (I am not going to deal with jurisprudential theories such as natural law but am emphasizing a positivistic approach) derives from state action (legislation, judicial decisions, regulatory rulings and promulgations by other public bodies) and articulates public policy enforced by governmental authority. Law prescribes and proscribes human (individual and collective) action to allow societies to function with some degree of predictability and, hopefully, fairness and justice. Law, however, is an imperfect mechanism for defining "Good Company" behavior since it is an artifact of political processes characterized by asymmetric power and the necessity to arrive at compromise approaches. The legal process, moreover, cannot anticipate the virtually limitless array of issues and concerns occasioned by corporate actions and articulates past approaches to yesterday's problems In short, law often articulates the "lowest common denominator" of socially acceptable behavior.

- *Affinity group regulation* in the modern context, enables affinity groups of whatever nature – business entities, professional bodies, religious authorities, social organizations – to establish standards of behavior for their members without the intervention of public authority. In the business realm, affinity regulation is of long-standing as in the case of the medieval European craft guilds whose members were subject to rigid regulations concerning all aspects of the trade. At its inception, the Hanseatic League and its member Hanseatics "were well appreciated as honorable merchants who ensured quality and fought against unscrupulous practices" through vigorous affinity regulation by the membership. (Kurlansky, p. 141). Today, the so-called "ancient-professions" such as law and medicine are governed by affinity regulation. Membership in the professional bodies is mandatory and they control licensing and discipline. Accountants and other health care professionals are similarly governed. Today realtors, architects and, security dealers (National Association of Security Dealers) possess professional codes governing their membership. Affinity regulation, if seriously administered by appropriate governing bodies, can have a salutary effect on member firms.
- *Self regulation* involves the voluntary acceptance of standards established by non-governmental entities such as the International Chamber of Commerce and the Organization for Economic

Cooperation and Development (OECD) or pertaining to particular issues such as child labor, South African apartheid (Sullivan Principles), third world apparel manufacturing, fair trade coffee, "green" environmental policies and the like. Such standards can serve the useful function of establishing "base-lines" which enable firms in an industry to engage in socially more efficacious practices without experiencing competitive disadvantage. Companies can also "sign on" to such voluntary codes of conduct such as the Caux Roundtable's Statement of Principles, the United Nations' Global Compact, or the European Commissions' Green Paper, all of which articulate general precepts of corporate "best practices" but lack any enforcement mechanisms. A primary value of such general codes of desirable business behavior is that they are educative for companies, industries and, indeed, nations particularly in developing countries (e.g. Lebanon and Croatia) which, in their struggles for national survival typically have paid little attention to issues of socially responsible business behavior. Self-regulation makes a difference only to the extent that companies adhere to the philosophical precepts and operative standards enunciated in these codes. Many companies, particularly in the United States and Europe, have developed Codes of Conduct or Ethics Statements as forms of self-regulation or have issued "Social Audits" reporting on their environmental, employment, and other practices.

- *Ethical precepts* derived from religious traditions, humanistic philosophy, customs, mores and traditions provide principles of "right action" with regard to human interactions both individually and in organizational settings. If law, as I suggested above, often defines the base-line or lowest common denominator of acceptable corporate behavior, ethics functions on a higher plane articulating standards of conduct which "go beyond" the legally required. Law and ethics are not, of course, mutually exclusive and legal requirements frequently are derived from and incorporate ethical precepts. Indeed, what were ethical aspirations for business behavior in one generation frequently become legal requirements in the next. Customs, traditions and mores can result, or course, in behaviors which are patently unethical such as "honor killings" of young women who are deemed to have violated traditional norms which, however, are contrary to civil law. We shall return to the role of ethics when we examine more closely the concept of corporate social responsibility.

- *Vigilant and responsible media.* In an open, democratic society, the media plays a critical role in informing the public about and rendering accountable all institutions which possess power. Indeed, it is the media, which often are the first source of information about governmental and business malfeasance or questionable behavior. One need only think of the all-encompassing Enron scandal or disclosures relating to dubious marketing practices on the part of major pharmaceutical companies to appreciate media's role in ferreting out illegal or unethical corporate behaviors. Business firms naturally do not welcome negative publicity which affect company reputations, incur governmental Interventions, generate legal liability, and may result in loss of business. Conversely, favorable media coverage can be politically useful, enhance business activity and stimulate socially responsible business behavior.
- Finally, *civil society* – An informed and engaged citizenry – is another important catalyst to encourage "Good Companies." This can results from direct citizens action in the case of consumer product companies – Nike and Starbucks are cases in point – or industries (e.g. agribusiness firms) by putting pressure on governmental officials to take action. One can point to Sarbanes-Oxley as an example of recent governmental regulation, which resulted, as least in part, from citizen concerns regarding corporate governance and transparency. Wal-Mart's efforts to move into several San Francisco Bay Area communities have been thwarted by citizen opposition within affected the communities typically acting through local government. To be sure, citizen action is not always effective. It does, however, have an important role to play in impacting the actions of corporations and those who manage them.

All six factors discussed above, individually and in combination are critical to achieving socially responsible corporate behavior.

3. The Good Company, Business Ethics and CSR

Let us turn specifically to the quest for the "Good Company." My basic thesis is that ethical teachings and "corporate social responsibility" although necessary, even essential, factors in achieving "good companies," are inherently insufficient. In descending order of importance, law, affinity regulation, self-regulation, the media and active civil society all play critical roles in achieving socially beneficial corporate behavior. In Western societies, this has always been the case and, if anything, is even more so today given the relentlessly increasing complexity of our twenty-first century global political-economy. My thesis is not based on a concept of human beings as inherently malevolent or evil but rather recognition of structural aspects of the contemporary business environment as well as the pressures on individual corporate managers. Among these aspects and pressures are: global competitive setting within which transnational corporations function; the emergence of a world-wide labor market made possible by technologies which have facilitated the fungibility of job venues; the appearance of truly global capital markets (consider the proposed merger for the NYSE and Euronext) which subject managers, irrespective of national boundaries, to short-terms accountability for "return on investment" and to unremitting expectations by financial intermediaries for ever-increasing profits; oftentimes, conflicting cultural norms which contribute to conflicting concepts regarding what constitutes "correct" corporate behavior; and, finally, an ideology of "profit maximization" which subordinates all other considerations to the Holy Grail of maximizing shareholder value. Putting aside managerial avarice as reflected in over reaching salaries and benefits at the top rungs of the corporate ladder, temptations exist in business organizations (and I include in this term, state owned or public enterprises where behaviors are often worse than in private enterprises) for managers to "cut corners" with regard to the environment, employees, producers, communities, third-world nations, and other stakeholders impacted by their policies and operations.

Before we briefly examine each of these factors, a final observation is necessary. Companies are multifaceted entities. As is the case with human beings, no company is 100% "good" and very few are 100%

"bad." In its ranking of the "100 Best Corporate Citizens 2006," *Business Ethics* utilizes eight measures for evaluating companies (community, corporate governance, diversity, employee relations, environment, human rights, product and total return). (Raths, pp. 20–28). Yet other criteria would be appropriate e.g. marketing practices and company philanthropy. Firms may be exemplary in their environmental practices and abominable in their employee policies. Similarly, companies may produce top quality, state of the art goods and services but be shady in how they market them. Moreover, practices may differ within the same organization based upon location, competitive position of a particular unit and the "character" (in both senses of the term) of corporate leadership. "Bad apples" exist in every company although, typically, unsavory corporate behavior results from organizational rather than individual failings. In short, business organizations are not monolithic in nature but have the capacity to manifest "the good, the bad and the ugly."

Let us look now at each of the above-mentioned elements:

- *Transnationalism/competition*. Transnational corporations function in a highly diverse globalized competitive world. The traditional productive factors of land, labor and capital, together with their modern addition, intellectual property, flow freely across borders or in the case of land (which also includes the air and sea) the natural products thereof (i.e. minerals, timber, agricultural items, fish, wildlife and fossil fuels). We function in a world of global markets for all factors. Business firms seek to acquire or utilize these productive resources as cheaply as possible irrespective of the impact of their operations on individuals or communities. Jobs are out-sourced to low labor cost centers notwithstanding the effect on home communities long dependent on these companies for employment. Natural resources are bought or leased and removed from less-developed countries for use in wealthier nations with little consideration given to environmental, social and economic impacts, including distributive effects, on the poorer region. Many companies adhere, at best, to the minimalist (if any) regulatory standards extant in many parts of Africa, Asia and Latin America rather than the "best practices" which they employ in more demanding home environments.

To be sure, where transnational companies invest in host regions, employ "best practices," respect human rights, develop an educated labor force, safeguard the environment, protect employees' health and are not complicitous with corrupt authoritarian regimes, they are forces for good in the development of these areas. Where, however, they pursue policies which are exploitive of less developed nations, they contribute to the growing economic divide between northern and southern hemisphere nations with deleterious effects on regional and global peace. In theory, free trade is intended to benefit all participate in the global trading order. To date, of course, benefits have been asymmetrical benefiting wealthier nations and contributing to interstate and intra-state conflicts.

- The emergence of truly *global financial markets* and the Anglo-Americanization of these markets have contributed greatly to the short-term perspective which drives much of corporate activity today. Although in theory, managers are supposed to focus on an organization's long term mission and goals, de facto both from within the organization and externally, they are subjected to unremitting pressures from large financial intermediaries such as mutual funds to concentrate on shorter term (monthly, quarterly, annual) financial objectives, lest their performance fall short of market expectations and thereby decrease the capital value of the firms and their own standing and financial position within the organization. While short-term expectations and pressures were at one time primarily concentrated in the United States and later the United Kingdom, they have now spread to many parts of Europe and Japan. Companies whose financial performance diminishes the value of their stock face, moreover, the risk of unfriendly acquisition. All told, these pressures often result in managers to becoming unifocal in their to attempt to maximize organizational profits by any means possible to the detriment of other considerations.
- *Defining the "Good Company."* Another factor complicating life for transnational corporations functioning in highly diverse global settings is that at times it is quite difficult to determine precisely what it means to be a "good" company. Although global codes of conduct

such as the Universal Declaration on Human Rights, the O.E.C.D. Guidelines for Multi-National Enterprises, the UN Global Compact, and the European Commission "Green Paper" on Social Responsibility, articulate desirable norms for corporate behavior, their provisions are often quite general and have no enforcement mechanisms. They are aspirational precepts rather than operational standards. As a practical matter, it often comes down to companies striving to achieve some balance between the legal requirements and informal expectations which prevail in their home countries and the legal requirements and norms which exist in the overseas setting. Managers perforce have to maneuver cautiously between conflicting definitions of human rights found in Saudi Arabia or Nigeria as opposed to the U.S. or the U.K. Definitions of nepotism and corruption are subject to regional variation. The line between adhering to universally accepted hyper-norms and engaging in "cultural imperialism" by imposing non-traditional standards in widely divergent societies is often quite thin. Accordingly, trade-offs exist for firms seeking to be "good" and "do good."

- Finally, there is the matter of *ideology*. Market capitalism has been nurtured on the concept that the common good is maximized when individuals and, by implication, organizations pursue their parochial self-interest. This perspective is attributed to Adam Smith (Smith, *The Wealth of Nations*) but in fact is only part of Smith's overall message that individual well-being can only occur where there exists underlying "sympathy" or recognition of a common good (Smith, *The Theory of Moral Sentiments*). Otherwise, we would live in an anarchial Hobbisian society – "nasty, brutish and short" – to the detriment of all. This broader Smithian perspective views self-interest as appropriate but an approach which must be "enlightened" by a broader perspective. Unfortunately, this more capacious part of Smith's message has been lost upon many business theorists and practitioners. The purported interests of the shareholder is deemed to trump all other considerations, a least in theory, although there is ample evidence that this objective is not adhered to in managerial practice.

The notion that business organizations have obligations to other stakeholders and, more generally, constitute a "public trust" has received little attention until quite recently. This fact is paradoxal given the fact that the earliest corporations were established to fulfill public service objectives: to build turnpikes, railroads and canals; to provide reliable sources of credit through state-chartered banks and to stimulate the growth of essential industries. How ironic it is that we are only now returning to recognition of these societal objectives of corporate enterprise.

4. Achieving "Good Companies"

So where all these constraining factors leave us in our efforts to achieve the "Good Company"? I have suggested above that in the corporate context these are six modes of social control of business behavior: law, affinity regulation, self-regulation, ethics/CSR the media and concerned civil society. I have also argued that none of these mechanisms is sufficient unto itself and that they are mutually reinforcing. To illustrate, media attention on deleterious products (e.g. flammable children's wear) can catalyze citizen concern about corporate "irresponsibility" that arouses governmental regulatory action via legal processes (legislation, litigation, rule making) or alternatively, affinity (children's wear industry) regulation or self-regulation by adopting a company code which prohibits such fabrics undertaken to pre-empt legal action.

What is the role of our concepts of ethics and corporate social responsibility in stimulating socially beneficial corporate behavior? Are they irrelevant concepts? Of course not! They have contributed to ever-escalating standards for corporate performance. Business practices commonplace in the mid twentieth century – practices relating, for example, to the environment, industrial health and safety, employment policies – are unthinkable in many, albeit not all, post-industrial societies. Concepts of global ethical standards – "hyper" or transcendent norms (Donaldson and Dunfee 1999) – relating to human rights, sustainable environmental policies, and employment practices relating to women and children appear to be having an impact in some less developed countries. It has been suggested that the U.N. Global Compact – although voluntarily, quite general in character and lacking enforcement mechanisms – and the European Commission's Green Paper relating to Corporate Responsibility have heightened awareness of and stimulated serious discussion about socially responsible business behavior in countries where this concept was dormant at best or even unknown.

Years ago, I argued it is useful to think about CSR in terms of both product and process (Epstein 1979). The product approach considers social responsibility in terms of outcomes – whether management has done the "right thing" or has achieve a recognizably "good result" by making the "right" decision about, for example, installing air quality safeguards beyond what is required by law. I also suggested corporate social responsibility could also be usefully thought of as a process: a system of decision making whereby corporate managers seek to anticipate and consider the total consequences of their companies policies and operations before they act. This product/process view of CSR, of course, goes significantly beyond the notion of social responsibility being simply synonymous with "good works" – e.g. financial or in-kind contributions to the community or Red Cross. Subsequently, I expanded the product/process approach to what I termed "The Corporate Social Policy Process," (Epstein 1987) which is delineated in Figure 1.

I maintain the belief that the Corporate Social Policy Process concept which integrates essential elements of traditional notions of business ethics, corporate social responsibility and corporate social responsiveness is useful in assisting managers who desire to have their organization be good "corporate citizens" (to use a currently popular term) to think through the implications of company decisions in a more thoughtful/analytical manner.

The argument which underlies this paper is that while we must maintain, indeed, invigorate efforts to articulate and implement ethical precepts and concepts of CSR derived from multiple sources to influence and enable managers and their organizations to act in a socially beneficial manner, we must recognize that there are limitations in terms of what we can accomplish by reliance on ethics and CSR.

I subscribe fully to the analysis of my Berkeley colleague, David Vogel, in his recent book *The Market for Virtue: The Potential and Limits of Corporate Social Responsibility* (Vogel 2005) wherein he points out that

Figure 1
The Corporate Social Policy Process
Contributions of Business Ethics, Corporate Social Responsibility, and Corporate Social Responsiveness to the Corporate Social Policy Process

Corporate Social Policy Process

Institutionalization within the corporation of processes facilitating value-based individual and organizational reflection and choice regarding the moral significance of personal and corporate action. Individual and collective examination of the likely overall consequences of such actions, thereby enabling the firm's leaders both individually and collectively within the organizational setting to anticipate, respond to and manage dynamically evolving claims and expectations of internal and external stakeholders concerning the products (specific issues or problem-related consequences) of organizational policies and behavior.

=

Business Ethics

Value-based ***reflection*** and choice concerning the ***moral significance*** of individual and organizational action by business decision makers. This reflection and choice emanates from and pertains to critical issues and problems confronting the organization and its leaders.

+

Corporate Social Responsibility

Discernment of specific issues, problems, expectations and claims upon business organizations and their leaders regarding the consequences of organizational policies and behavior on both internal and external stakeholders. The focus is upon the ***products*** or corporate actions.

+

Corporate Social Responsiveness

Development of individual and organizational ***processes*** for determining, implementing and evaluating the firm's capacity to anticipate, respond and manage the issues and problems arising from the diverse claims and expectations of internal and external stakeholders.

Source: Epstein, 1987, p. 107.

notwithstanding the best of intentions, there are constraints on the ability of companies and their managers to achieve sociably responsible behaviors. Consistent with the argument which I have made above, he assents: "The important complementary relationship between civil and government regulation suggests that the definition of a responsible corporation needs to be expanded. Corporate responsibility should be about more than going "beyond compliance;" it must also include efforts to raise compliance standards. In fact, the most critical dimension of corporate responsibility may well be a company's impact on public policy." (Vogel 129)

Raising the bar of what constitutes appropriate corporate behavior through governmental regulation enables firms wish to adhere to higher standards of business behavior to compete on a more level placing field. Additionally, truly effective affinity regulation can play a complementary reinforcing role.

To summarize, we must, accordingly, rely on legal processes, both nationally and trans-nationally via regional and global institutions, to establish the "rules of game "for desired corporate behaviors. In addition, we must encourage business organizations to engage in meaningful, as opposed to cosmetic, affiliate and self-regulation of their activities on an industry or regional basis and thereby obviate the necessity for additional public policy initiatives. Moreover, ethical norms and aspects of CSR embodied within the concept of the corporate social policy process will assist managers in assessing the broader societal implications of their decisions. And finally, vigilant media monitoring and vigorous citizen oversight of corporate action are essential to assure that business firms are both rendered accountable for the consequences of their actions and exposed to public expectations of "best practices" of business performance.

Perhaps a more realistic goal as we think about the desire to stimulate the spread of "Good Companies" which connotes the achievement of a certain type of corporate perfection is to strive for "Better Companies," organizations which continuously seek to perform the economic functions for which society relies upon them in a manner which optimizes the firm's utility to the diverse stakeholders affected by their actions and minimizes the deleterious effects (what economist term "externalities") of their operations.

References

Donaldson, T. & Dunfee, T. W. (1999). *Ties That Bind: A Social Contract Approach to Business Ethics*. Boston: Harvard Business School Press.

Epstein, E. M. (1979). Societal, Management and Legal Perspectives on Corporate Social Responsibility – Product and Process. *Hastings Law Journal* 30: 5, May 1979, pp. 1287–1320.

Epstein, E. M. (1987). The Corporate Social Policy Process: Beyond Business Ethics, Corporate Responsibility and Corporate Social Responsiveness. *California Management Review*, 29: 3, Spring 1987, pp. 99–114.

Kurlansky, M. (2002). *Salt: A World History*. London: Vintage Books.

Raths, D. (2006). The 100 Best Corporate Citizens 2006. *Business Ethics*, Spring 2006, pp. 20–28.

Yergin, D. & Stanislaw, J. (1998). *The Commanding Heights*. New York: Simon and Schuster.

Smith, A. (1966). *The Theory of Moral Sentiments*. New York: August J. Kelley Publishers.

Smith, A. (1937). *The Wealth of Nations*. New York: Modern Library.

Vogel, D. (2005). *The Market for Virtue: The Potential and Limits of Corporate Social Responsibility*. Washington, D.C.: Brookings Institution.

Imre Ungvari Zrinyi
"Babes-Bolyai" University Cluj

Dialogic Ethics for Business

1. The Need for a Dialogic Perspective

The search in philosophy for a meaningful wholeness, an all-embracing principle, which relates the reason of human endeavor with the universal basis and the order of things and meanings, was a common trait of several metaphysical conceptions and also a matter of endless debates between philosophers with different convictions. Thus the idea of such a wholeness, or basic order has become for now subject to a justified suspicion, as a result of our experience in centuries of abusing the construed notions of generality. The basic problem of social cooperation and contemporary social philosophy is the interpretation of social relatedness, of the nature of social ties. Social research and common thinking equally lack a widely accepted term, which would be able to formulate the legitimate frame of social cooperation, while maintaining the freedom of action and selfrealization of different individuals and cultures. As contemporary philosopher, Michael Gardiner, stated, the disintegration of the cultural and ideological homogeneity characteristic of modernity has resulted in a relativizing of moral discourses, and a concomitant proliferation of values, lifestyles and worldviews.

The joining of opposite traditions, the realization of relatedness and joint interpretation can be assured only by communication. However we are strongly committed to our worldview and system of values by our cultural tradition and education, we cannot expect that all the others be willing to share it with us; we must accept that the encounter of two people possibly gives life to a new view. If this does not happen, both of them miss the encounter and a considerable part of possibilities with it. What appears to be an obvious starting point is that: *if we want to take advantages not only from the manifoldness of relations but the possibilities resulting from the multiplicity of interpretations as well, then we must choose those forms of cooperation that take into account the participants' interpretation of reality.* This statement is valid for any cooperation in any group where the outcome of the activity is inherently connected to the creative contribution of each participant.

As a result thinkers who were inspired to realize the dialogic nature of human existence, the possibility of surpassing the distance between the Self and the Other has become of major importance. They discuss how prejudices can be defeated and how dialogue leads to mutually advantageous decisions. Regardless of the differences between their basic ideas, we will call 'dialogic' all those theories that consider the surpassing of the monological standpoints a basic problem (e.g. the work of Martin Buber, Emmanuel Levinas, Hans-Georg Gadamer, Karl-Otto Apel and Jürgen Habermas).

2. Philosophical Conceptions of Dialogism

Dialogue may be taken as a keyword of our era, which rules philosophy, theology, ethics, social and political theories in an unquestionable way, just like "substance" or "reason" in the earlier times. As George Myerson states in his book about rationality as dialogue, many thinkers now consider reason as an aptitude to lead a dialogue, to obtain a communicative agreement. In the same time, dialogue became important not as a new form of reason, but as the basic word for openness and attention directed toward the other's nature and original otherness. That is to say, the principle of *dialogism* is not a new term for rationality, rather the fundamental category of being-in-between. According to Emmanuel Levinas: "The value associated with the concept, the practice and the word dialogue (that is, the discourse which takes place between people in face-to-face situation, interpellating and changing statements and objections, questions and answers) by a whole range of philosophers, theologists, ethicists, politicians and even public opinion, attest the new orientation of the idea which Western society creates for himself about the essence of the rational and the spiritual as a result of those trials which 20th century has undergone after the first World War" (Levinas 1980).

The real importance of the philosophies of dialogue consists in the transcendence of the ego's isolation and the instrumentalist character of their connections by emphasizing the elements of encounter, the recognition of the otherness and the cooperation in carrying into effect their interests. The development of such a sensitivity constitutes the real stake of the philosophy of dialogue. In the next passages we are trying to present some examples of dialogic philosophy with the intention to display their possibilities for the renewal of both the human social relations and moral thinking.

2.1 Martin Buber's 'I'–'Thou' Relation

The philosophy of dialogue in the narrow sense of the word is the philosophy of Martin Buber expressed in his wellknown study *'I' and 'Thou'* (1923). Buber's conception of dialogue is deeply rooted in Jewish spirituality, especially in the religious movement of Chassidism. The Chassidic faith that God takes his dwelling-place within man and therefore – regardless of the infinite abyss between them – the dialogue of man and God is possible, was determinant for Buber's idea of dialogue. Buberian dialogism views the human existence in relations and extends the relations over interhuman and God–man relations to the whole of existence. From the point of view of our inquiry that looks for the elements of a genuine (ethical) dialogic attitude it is of a crucial importance Buber's distinction between two fundamentally different kinds of relations: the I–It and I–Thou relations.

An I–It relation is the normal everyday relation of a human being towards the things surrounding him. In I–It relation the self confronts an external object-world, and proceeds to give this world shape, meaning and a pragmatic 'use-value'. The I–It relation is one of detachment and mastery, and regulated by purely technical interest in manipulation and control. The I–Thou relation is conceived as radically different. In this relation "I" appears as a person and becomes conscious of itself as subjectivity. The human being enters into it with his innermost and whole being, in a meeting, in a real dialogue – this is what both of the partners do. Therefore the relation to the Thou is immediate. Between I and Thou there is no terminology, no preconception and no imagination, there is no purpose, no greed and no anticipation; All means are impediment. Only where all means fall to pieces, encounter happens. In this sense in the I–Thou relation we find the elements of a real intersubjectivity.

In Michael Gardiner's approach "The I–Thou is a qualitatively different type of relation since it establishes a new type of entity, a 'We', which denotes a subjective, highly personalistic relation between two flesh and blood human beings located firmly in actual time and space. When it enters the I–Thou relation, the self comes to the realization that it cannot be a self-constituting, autonomous ego, but part of the category 'in between', or what Buber sometimes calls 'the ontology of the inter-human'." (Gardiner 1996: 125). Dialogue is rooted in everyday life, and the responsibility the I–Thou embodies is in reality a response to the concrete demands of daily social existence.

2.2 Levinas and Sensibility for Otherness

In Levinas' view reality is complex and internally divided. The subject itself is not a unified entity, but radically pluralistic and it is generated out of the alterity between the self and the Other. An understanding of this alterity requires that we cultivate a sense of 'exteriority', and develop an awareness of our relationality to the world and other selves. The simple *presence* of the Other implies a situation of addressivity, in which I am summoned by the Other in a manner that demands a response from the self. One cannot remain a neutral bystander, indifferent to the presence of the Other. To answer the summons means the acceptance of responsibility. So the Other is not conceived of as object; but nor is there any desire to ecstatically fuse with the Thou. In the encounter there is, along with communion, a maintenance of distance, by which we return to our own self. We do not dissolve ourselves in the Other, as in the psychological phenomenon of empathy or *Einfühlung*.

For Levinas reciprocity is an inadequate metaphor to capture the ethical quality of the self–Other relation. The whole logic of reciprocity is mired in contractual, utilitarian thought and hence tainted by egoism (Levinas 1980: 131). By contrast Levinas wants to argue that what is noteworthy about the self–Other relation is precisely its asymmetrical and non-reciprocal character. The essence of dialogue is that it demands a response – not for *what* was said, its semantic content or descriptive elements, but in terms of the nature of the relation it forges. This implies responsibility, but not a responsibility that is contingent upon reciprocity or justice. Responsibility to the Other pre-exists self-consciousness; it bypasses rational, calculative thought. I do not grasp the Other so as to dominate, but I respond to the face's epiphany as if to a summon that cannot be ignored. Levinas asserts that this unconditional responsibility is announced in the statement 'Here I am', which announces my essential 'openness' to the Other. Responding to the Other means that we must respond as a *unique* being, for whom no one else can act as a substitute.

The key to understanding Levinas's construal of the relation between alterity and ethics lies in his notion of *proximity*. Proximity is 'not a coincidence or a lost union', he suggests, 'but signifies all the surplus or all the *goodness* of an original sociality'. Insofar as I and Other must share proximity, my being-in-the-world cannot be a usurpation of the Other's right to be.

2.3 Gadamer's Concept of Understanding through Dialogue

Taking Heidegger's ontological hermeneutics as a point of departure, Hans-Georg Gadamer negates the idea of a wholly autonomous subject which in his cross-historical and cross-cultural text-interpretations is able to transcend an already existing sense by his own. Instead of such a timeless, monological interpretation Gadamer develops a dialogical model of understanding, in which the text is a "thou" whom we are engaged in conversation. The task of interpretation is to make something alien accessible. Hermeneutics emphasizes the role of the tradition in constituting any present or future understanding. Therefore understanding is not a pure act of consciousness, but it is conditioned by a perspective including affections, concepts and practices of cultural heritage constituting the interpreter's horizon. A perspective is not a "closed" circle of understanding, but is rather, by definition, an open and moving set of directions of pursuit. Like any experience, cultural experience is a directed movement towards expanding interpretation, a "fusion of horizons" that meets other interpretations as partners in an interpreting dialogue.

Gadamer uses dialogue as a model for interpretation in general. To speak "with" another does not require that the other conform to our opinions, nor to "put ourselves into someone else's shoes"; it rather takes the form of mutual orientation toward a single topic. To interpret another requires that the interpreter takes her text (person, book, or culture) as a living, self-questioning, open-ended system of actual possible contributions.

Gadamer takes the Platonic dialogues and Aristotle's notion of practical judgment or *phronesis* to be exemplary for hermeneutic practice. In Gadamer's dialogical model *phronesis* is the non-methodological application of general principles to particular situations. While moral principles, like theoretical principles, can be taught, the application of moral principles is a matter of character, formed by cultural and social habituation (*ethos*). In this way, Gadamer views the dialogue as an application of traditional habituations (prejudices) to new and unpredictable situations.

2.4 The Communicative Ethics of K.O. Apel and J. Habermas

Communicative or discourse ethics is a specific type of ethical theory worked out by Karl-Otto Apel and Jürgen Habermas; its central insight is the argumentative justification of the universalizability of norms. According to the founders of this perspective only those norms and normative institutional arrangements are valid, which individuals can or would freely consent to as a result of engaging in certain argumentative practices. Such argumentative practices are described by Apel as "an ideal community of communication", while Habermas calls them "practical discourses". Both agree that such practices are the only plausible procedure in the light of which we can think of the Kantian principle of "universalizability" in ethics today.

Universalizability is defined as an intersubjective procedure of argumentation, geared to attain communicative agreement. There is also a shift from the model of the goal-oriented or strategic action of a single agent intending a specific outcome to the model of communicative action which is speech and action to be shared with others. The conversational model that can be able to avoid the charges of dogmatism and circularity is named by Habermas the "ideal speech situation" and is considered as "universal and necessary communicative presuppositions of argumentative speech" entails strong ethical assumptions. These assumptions geared to require of us that we recognize the right of all beings capable of speech and action to be participants in the moral conversation. These conditions further stipulate that within such conversations each has the same symmetrical rights to various speech acts, to initiate new topics, to ask for reflections about the presuppositions of the conversation, etc. The very presuppositions of the argumentation situation then have a normative content that precedes the moral argument itself. Such a presupposition is for Apel the principle that all beings capable of speech and action are potential members of the same communication community with me, and that they deserve equal and symmetrical treatment.

As a result of the perspectives of the philosophies of dialogue that it is possible to develop out genuine ethicality on the base of a free, untrammeled dialogue between the self and the Other. In such a dialogic encounter, the rigid boundaries of the egoistic, hyper-rational self are breached, and this self garners a renewed awareness of, and respect for, otherness. Otherness is no longer foreign and threatening. Dialogue, thus establishes, becomes an end to itself, inasmuch as it establishes a relation of mutuality, shared responsibility (or 'answerability') and unsolicited concern. Yet at the same time the autonomy of the interlocutors is ultimately preserved – there is a synthesis of communion and distance in genuine dialogue. The integrity of the dialogic encounter is contingent upon its embeddedness in everyday sociality – in communicative and practical forms of intersubjectivity that constitute the connective tissue of the lifeworld (Gardiner 1996: 140).

2.5 The Dialogic Self and Culture as Communication

The new orientations and major discoveries in the realm of philosophy, mainly in the field of phenomenology, semiotics and hermeneutics have had a determinant impact on the development of human and social sciences in the postmodern era. Due to this impact the human sciences have been retreating from relaying on metaphors drawn from the more established physical sciences to adopting metaphors drawn from history, biography, literature, drama, and rhetoric (Sarbin 1993: XIII). As a result of this conceptual turn there appear in philosophy and psychology new useful insights, such as 'narrative identity', 'dialogical self', 'conversational behavior' etc., which emphasize once again the special relevance of dialogue for the study of social behavior and especially for ethics.

Alasdair MacIntyre in his *After Virtue* claims that the sense of human actions (as vell as the sense of an utterance) becomes intelligible by finding its place in a narrative that provides him a context in and by reference to which purposes and speech acts are rendered intelligible. He also states that the most familiar type of such a context is the conversation regarded as a dramatic work in which the participants – equally actors and joint authors – work out in agreement or disagreement the mode of their production. Conversation in this sense becomes a basic category of interpretation of acts with major social and ethical implications: "conversation, understood widely enough, is the form of human transactions in general. Conversational behavior is not a special sort or aspect of human behavior, even though the forms of language using and of human life are such that the deeds of others speak for them as much as do their words. For that is possible only because they are the deeds of those who have words" (MacIntyre 1981: 211).

The moral relevance of widely understood conversation in MacIntyre's view seems to be obvious: persons gain their moral quality through their accountability by their participation in the correlated, interpenetrating narratives. To be subject of a narrative comprising the whole life means to be accountable for the actions and experiences involved in it, but also to be able to ask others for an account. So personal identity, narrative, intelligibility and accountability are bounded in an inseparable manner. One of MacIntyre's central thesis is: "man is in his actions and practice, as well as in his fictions, essentially a storytelling animal." The storytelling as a form of social relatedness and interaction is basic for any narrative discipline.

Similarly, the 'dialogical self' in Hermans' and Kempen's narrative psychology makes sense in the context of story making and storytelling. The conception of the dialogical self helps to make clear and to shape a central feature of narrative psychology: the collaborative nature of self-narratives. In the way Theodore R. Sarbin understands the significance of the 'dialogical self': "Contrary to the implications of a monological view of self, the dialogical self is formed out of the construction and reconstruction of encounters with others and the reciprocal influences of multiple others."(Sarbin 1993). As we have seen, dialogism, or the narrative–communicative conception of human social condition tends to become a new, unifying approach of the entire cultural life. An appropriate field for work on such a unifying attempt seems to be cultural anthropology.

For the anthropologist Nigel Rapport, who relays on the theories of A. MacIntyre, C. Geertz and M. Oakeshott, conversation, stories which can be told together constitute the essential contexts of social coexistence. In his conception Oakeshott's idea according to whom "conversation is a meeting of voices speaking in different idiom or mode" becomes more extensive and through the theory of speech modes embraces the whole sphere of culture. In this respect science, poetry, practical activity, history, symbol systems, are modes of speech, different universes of discourse. Going on in public and inside themselves, the conversation ultimately contextualizes every human activity and utterance (Rapport 1997: 178).

3. Towards a Dialogic Attitude in Business Activity

3.1 Dialogic or Monologic Conception of Business?

The dialogic viewpoint corresponds to a necessary change of views in business activity. In a world of global economy, which means large scale and continuous changes in economic, social and legal environment, business faces an unprecedented complexity of its internal and external relationships. The increasingly competitive world market, the democratic institutional and legal system with its strict business regulations, well-organized global civic movements, more informed consumers and more innovative competitors require a more responsible attitude, a quickly reacting ability, more flexible organizational structures, with greater emphasis on communication and self-organizing competence. As John R. Allert remarks, working in a transnational world presupposes communicating across differing political socio-economical and cultural dimensions with the relatively new and rapidly changing tools of information technology, and at the same time, abiding within a range of accepted cultural ethical frameworks and communication styles (Allert 1999)

Due to the major changes in the legal system which empower persons and organizations to protect their interests against the abuses of corporations, and moreover the increasing expectations for business social responsibility, including its contribution to the stakeholder's and the society's overall welfare, the corporations' concern for consumers, employees and for the environment have gradually become part of the corporations' policy. So dialogism must be an alternative approach to a mere costbenefit analysis by introducing a new concept of efficiency, that of answerability, or rationality through dialogue, that requires to take account of others. Such a new systematic approach of business management is provided by the *stakeholder theory* of the modern corporation worked out by Edward Freeman.

In Freeman's opinion the task of management is to consider and balance not only the interests of the stockholders or owners, but the multiple claims of conflicting stakeholders such as *stockholders* or *owners, suppliers, resellers, lenders* of capital, *government agencies, political groups, employees* and *unions*, *consumers*, *competitors*, the *local community*, *society at large*, the *environment* etc. In the stakeholder theory vision the corporation ought to be governed in accordance with three sorts of principles: the principles of fair contract, the principles of caring, connection and relationship and the principles of caring for the earth. Each stakeholder must be free to enter agreements resulting contracts with clearly defined entry, exit and renegotiation conditions. The procedure for changing the rules of the game must be agreed upon by unanimous consent and so on (Freeman 1994).

The dialogic element of this conception became obvious in comparison with Milton Friedman's more orthodox conception of business. As it is well known, in the social responsibility of business debate Friedman argues for a business without any social responsibility except that of profit maximizing. So in terms of communicative relations his position is basically a monologic attempt to gain the reconciliation of divergent values through exchange, where all the participants pursue their different interests. Friedman express the basic assumption of his attitude using the metaphor of free speech: "the freedom to speak does not imply having an audience, just as the freedom to sell does not mean having a buyer, only the opportunity to seek one. Only if speaker and listener can mutually benefit will transactions take place in the marketplace of ideas" (Friedman 1987: 6). In the case of the stakeholder theory we find a real disposition to responsibility through dialogue. For those who are involved in the transactions or in their consequences 'freedom to speak' implies 'the right to be heard and taken into account'. In ethical terms the (affected) existence requires responsibility and care for its stake including a real bargaining with and/or consideration of his interests. In the course of bargaining with all affected parties the initially stated principles must be completed with new ones, with the rules mutually accepted during the process of agreeing, and all this with regard to the situation and the specificity of values.

As a result the contemporary morally engaged transnational industry will need to develop its ability to respond effectively to all the divers and conflicting forces at one and the same time, so as to effect efficiency, responsiveness and innovation without trading off anyone for the other. In other terms corporations today need a more efficient and morally more sensitive communication with all their stakeholders. Responding to these complex issues is largely the responsibility of the professional communicator and relationship manager: the public relations professional (Allert 1999: 189). This implies a new conception about corporation and its strategic decision makers.

3.2 The Functions of Communication in the Management of the Corporation

The possible consequences of bargaining, dialogic attitudes that appear in the stakeholder theory, are made relevant for a new, community creating spirituality by Peter Pruzan and Ole Thyssen, who evolved the dialogic concept of "corporate culture" (Pruzan and Thyssen 1990). The central concept of their view is the "value communication" of decision makers and stakeholders, the organizational culture based on dialogue, which creates the possibility of managing constructively in the field of corporation management the complexity, specialization and pluralism characteristic to postindustrial society.

Corporation, according to Pruzan and Thyssen, is a cultural entity, an autopoietic social system organizing in the value communication of stakeholders, which is in command of its own language comprising a set of themes and interpretations, and owns the capability of ethical attitude and consensus oriented problem solving. Consensus may be worked out on the level of individual decisions, on the level of rules and procedures regarding the differences in opinion, and on the level of values and preferences. In the course of solving conflicts, decision making must always be capable of transcending the hierarchic division of power and the formal rules, in order to develop a common horizon of values of the organization and to take advantage of complexity and specialization. This means that conflicts must be solved regularly on individual, group and corporation level as well, through autopoietic decision processes, where consensus is but a station in the flow of communication. Consequently, the corporate culture implies rather a spirituality worked out together than an ideology or conception imposed by the management. The idea of 'corporate culture' is the starting point for

Peter Pruzan to develop his conception of the value based management. "Value based management – as he states – presupposes that the organization and its stakeholders develop a shared language and tools that can help the organization to observe and reflect upon itself; to measure the extent to which it contributes to its stakeholders' value; and to make choices that promote the interests of the organization as a whole" (Pruzan 1999: 271). A good business ethics policy, which considers the stakeholder's and the society's overall welfare, and in which words and deeds agree, is the first line of defense against unethical or illegal activities; it can prevent fraud; it can motivate employees to be responsible, creative and faithful.

3.3 The Hard Tasks of Communication

In the establishment of the dialogic (communicative) organizational attitude, it is extremely difficult to make it accept that the corporation be lead by its own collectively established spirituality and its communication, and therefore management becomes only part of the organizational system and gains its power only by taking part competently in the conversations and debates within the system. This relation seems to be essential for an integrity based relationship. "If leaders participate in these relationships of integrity – writes Marwin T. Brown –, they can enjoy the same quality of human relationships as other members. One could argue, in fact, that leaders must be members before they can be leaders" (Brown 2006: 17).[1] In the meantime leaders are defined by their communicative and problem solving competence. According to Guttorm Floistad (Floistad 1999), communication in contemporary organizations becomes an art, essential to social and cultural competence. As he states, leaders of business culture need to be teachers of relationships and wholeness, and in order to be teachers they should be realizers and speakers. Similarly, John R. Allert considers that the professional communicators must take their place among strategic decision makers: "The purpose of all public relations and organizational communication is to build, consolidate and manage relationships; this function cannot be successful if the creator and communications strategist is not participant in the strategic thinking of the decisions-makers – the dominant coalition. (…) The public relations communicator must be involved from the start of any contract or negotiation so that the unique and specialized aspects of the relationship can be integrally embedded in the communications function" (Allert 1999: 193).

By the mere "empowerment of public relations in the dominant coalition" (Allert), the ethical relationship-building communication cannot be considered as already atteined. As Tomlinson remarks, the predominant model of public relations in use today is monological. Case-study books abound with examples of situations in which public relations practitioners did not listen to their publics or communication partners (Tomlinson 2006). Employee involvement plans have been (largely) developed more to increase compliance, commitment and loyalty and they are limited more to application decisions than to broaden value debate and to increase innovation, while customer focus groups often function more to solicit information on taste and pricing to aid sales. According to a study written by Stanley Deetz and his partners "*Most representation forums are used by management to suppress or to diffuse conflict arising from stakeholder groups rather than to foster genuine conflict and debate for the sake of company improvement*" (Deetz et al. 1997: 217). All these procedures of the management constitute an attempt to reduce the complexity of decisions applying "systematically distorted communication", and ultimately to substitute valuecommunication with mere instrumental practice.

According to Tomlinson it is obvious that only the dialogic relationship management perspective for public relations uses communication to develop, maintain, grow, and nurture mutually beneficial relationships. Adopting Pearson's view he declares that "establishing and maintaining dialogical communication between a business organization and its publics is a precondition for ethical business practices" which is also true in relation with any other stakeholder group or person.

3.4 Misunderstandings, Ambiguities and Differences in Opinion

Value communication is an exceptionally difficult and relatively long process: knowing values and preferences means a hard task for verbal communication. Considering the cultural conditions of understanding and the plurality of communicative intentions and their message validity claims, agreement about the fundamental corporate values remains for a while within the boundaries of contingency, moreover, if we take into account the ever widening scale of misunderstandings, ambiguities and differences in opinion concerning the assumptions, it often seems to us that it is a mere illusion. Under these circumstances, the condition of conversation regarding values is that participants agree in the meaning of words. This really happens, namely in the form of 'meaning tests' that are woven into the conversation – participants intuitively check whether all the others use the same language as themselves or play the same cards as they do. If there is no accordance, the conversation changes its topics and becomes selfanalysing until a common denominator is found, or it simply stops. In order to avoid misunderstandings and possible traps of strategic communication, it is obvious the usefulness of the four types of assumptions for a non-violent communication set up by Habermas: the participants articulate themselves clearly, they tell the truth, the rules they accept are valid and they are sincere.

In an increasingly multicultural context the value-communication must surpass not only personal purposes but also several cultural boundaries that influence their sense-making capacities. In such a context business people face one another across the context barrier, making miscommunication inevitable. According to Janelle Brinker Dozier, Bryan W. Husted and J. Timothy McMahon "false assumptions regarding shared context may be one key to ethical misunderstandings between people from different cultures" (Brinker Dozier et al. 1998: 113), so even in everyday situations of truth-telling and promise-keeping must be understood in terms of the level of contexting in a given culture. But the authors also emphasize that the challenges of cross-cultural ethics can be ameliorated by realizing that some differences are due to differences in communication dynamics, rather than to differences in ethical standards. So differences in communication styles must be separated from true differences in ethical attitudes. Such difference of communication styles and dynamics were described by Edward T. Hall as "low context" and "high context" cultures. There are three major outcomes of communication that are closely related to the cultural level of contexting: (1) facts, (2) face saving and (3) harmonious interpersonal relationships. Low context cultures value facts and "hard data", that is, statistics, quantitative information and absolute, objective truths. In a high-context culture there is a "public truth" which preserves everyone's face, and the "real truth" which is understood by all, but not verbalized. Communicators from low-context cultures value facts, even if this undermines personal face and harmonious interpersonal relations for their more highly-contexted counterparts. Conversely, people in high-context cultures will sacrifice facts in order to save face and promote harmony (Brinker Dozier et al. 1998: 114, 115).

Although those differences seem to characterize nations and cultures, e.g. "anthropologists classify Anglo, Nordic, and Germanic cultures as low-context; Latin-European, Latin American, Arabic and Far Eastern cultures are classified as high-context." (Brinker Dozier et al. 1998: 113) If we take into account the professional origin of senior management members in different periods, as Allert does, we realize that even in the one and the same culture different professional origins will cause differences in communication styles. Searching for the place of the public relation specialist among the top management strategists, Allert asserts that in the historical context of corporate business senior management has largely come from those professions that value quantitative skills as precision, formula, measurement and evaluation, dealing with facts and figures. Conversely, as he states "the public relations practitioner… has been mainly concerned with intangibles, such as goodwill, corporate image, sense-making, relationship management and levels of satisfaction….These interpersonal 'soft skills' embrace the protocols of human interactions that rely heavily on trust, values, ethics and belief systems"(Allert 1999: 192). As a result, those professionals who are aware of their possibilities in approaches to cross-cultural communication differences will be able to surpass their cultural parochialism and choose consciously between different strategies as conformity, compromise, cultural parochialism and synergy. If they want to embed values and ethics in the communication function, then the strategy must be synergy, that consists of developing norms, policies and practices that are acceptable to both high- and low context people" and this – in the long run – will result in both better decisions and better relationships (Brinker Dozier et al.1998: 121)

3.5 Communication Styles and Conflict Resolution

As we presented before, business people are increasingly called upon to resolve different sorts of conflicts, including moral conflicts between various stakeholders who maintain opposing ethical positions or principles. The conflict solving capacity of different communicative styles and their relation to different conceptions of morality are now in the front of the business communication and business ethics studies.[2] French and Allbright have analyzed the tendencies of resolving moral conflict in business negotiations with various stakeholder groups which hold opposing moral beliefs and ethical standards. They realize that moral dilemmas could be easy turned into conflicts or in less adversarial reactions by differences in the communicative style used by each party in the dispute, e.g. strategic tactics such as intimidation, threats and the subordination of the opponent or their alternatives as communicative actions which encourage all affected parties to agree to arrive at a new position that takes into account the welfare of all those affected by the conflict (French and Allbright 1998: 177).

In spite of the fact that French and Albright's research results does not confirm Habermas' theory concerning the parallel relation between stages of moral development and the communicative behaviors, people exhibit in attempts to resolve conflict, they find some positive relationship between "win-lose" results and belonging to a morally lower level group than the higher level group" (French and Allbright 1998: 189) and they also found that when one party attempted to resolve the conflict through a discursive approach, the other party tended to react in a less adversarial manner. So they consider that "Habermas' theory of discourse ethics requires an operators' manual as well as a proactive implementation if it is to be of use to business" (French and Allbright 1998: 191).

In their opinion steps towards working out an operational version of the discourse ethics have been recently made by a project undertaken by the Levi Strauss company which put its managers through a training program structured around a communication process advanced by Stephen Toulmin and Marvin Brown. The Toulmin–Brown procedure incorporates the ethical goals of Habermas who requires that an agreement be reached on a mutually acceptable position. In fact, the starting point is to identify principles shared by both parties to the conflict and build upon those commonalities to construct a new position.

Taking into account both the philosophical and practical considerations, we are convinced that dialogic business attitude is a valid and desirable one, but it cannot be justified and maintained by mere economical arguments, nor by the considerations of a self-validating, absolutist moral code, or mutuality expressed in the golden rule. The ideal of dialogic attitude is inherently related to certain worldviews, to people's beliefs and their intentions to adopt a responsible position towards their social and natural environment. This idea emerges from their endeavor to recognize themselves as members of a selfmanaging moral community, which is able at the same time to articulate and sustain the requirements of a specific spirituality.

4. Conclusions

Dialogic conceptions focus on the relational and discursive origin of ethics rather than communal, individual or rational ones. As the philosophers of dialogism have demonstrated it, the normative core pursued may not be substantive, internalized values held by a society, but rather moral discursive procedures already practiced in community. According to Habermas moral procedural guidance for decision making is found in the immanent conditions of communicative action, in the communicative micro-practices of everyday life. So, we can state that responsibility is not only an element of an already existing specific social culture, but can be conceived as inherently linked to man's whole range of relations be they either relations between persons or relations between persons and other entities.

To draw an entity into the communication of interests presumes to look at the meaning and importance of the interaction not only from the narrow technical point of view of our own intentions, but from the entity's own articulated standpoint (in the case of persons) or – if they are not able to affirm it, like the nonhuman sphere of nature – we must interpret their well being according to the main tendency appearing in the relationship with the whole of their wide context. This relationship includes their participation in a communicative (transactional) process, and always requires a necessary openness to dialogue. Our assumption is and the above mentioned studies confirm that this idea could be extended in the realm of the ethical business activity and business communication.

Culture is not only an ongoing creation of the everyday activities of interconnected people, but also a system of senses that inherently functions as a constraint and valuation on what is socially enacted and understood. Communication ultimately contextualizes every human activity. To conceive business as a system of communication and negotiation through which different interests could enhance societal well-being as well as meeting the personal needs of the negotiating parties, means to place in the center of business activity the dialogic attitude. It is only too obvious that cooperation regarding the mutual benefit cannot be else than a cooperation that takes into account the interpretation of the other side. Consequently, moral relation necessarily comprises a hermeneutical component: this is explained by the dialogic idea of morality. Dialogism is a comprehensive conceptual frame, a certain vision of people's relatedness, the possibility of their agreement. Parties in a dialogue although they preserve their otherness and independence, accept the necessity of working out meaning together.

My concerns in taking dialogue as a model for an honest relation between persons in everyday life are the followings: (1) contemporary societies are subject to differentiation and interdependence so that any intention of a personal goal achievement in any domain, seen in its concreteness meets several contributions and obstacles from other intentions alike; so major purposes can no longer be substantiated monologically, they have become subjects of a dialogue; (2) living in the everyday life-world, people lead their life in a permanent contact and interaction between 'self ' and 'other', they must choose between the conflict and the dialogue of intentions and differences; (3) an essential trait of sociality – as the German philosopher Max Scheler stressed –, is exactly to be present 'for one another'; (4) dialogism provides us with a 'situated ethics' that represents an alternative to a coercive moral absolutism; (5) dialogic conceptions focus on the relational and discoursive origin of ethics rather than communal, individual or rational one; (6) by including dialogical conceptions in the ethical discussion about business activity we hope to reach an insightful, more creative, responsive and responsible praxis of management.

Notes

1 Brown's idea is another argument for leadership's integrity considered by Pruzan to be the most difficult trial in the establishment of corporate culture. "To establish a dialogueculture – writes Peter Pruzan – where the corporation can observe itself through the ongoing conversations that create and recreate the organization as an interplay between its stakeholders...that presuppose harmony between management's words and deeds. Values commit their 'user'. They demand consistency. People are compelled to act in accordance with their words, contrarily they risk losing the confidence and trust that are necessary preconditions for valuebased dialogue" (Pruzan 1999: 270).

2 See Tomlinson's study on monologic and dialogic communication, Brinker Dozier and her colleagues' study on communication approach to cross-cultural ethics, and French and Allbright's study on resolving a moral conflict through discourse.

References

Allert, J. L. (1999). Ethics in Communication:The Role of Public Relations. In S. K. Chakraborty & S. R. Chatterjee (Eds), *Applied Ethics in Management. Towards New Perspectives*. Berlin, Heidelberg, New York: Springer Verlag.

Benhabib, S. & Dallmayr, F. (Eds) (1991). *The Communicative Ethics Controversy.* Cambridge, Massachusetts, London: The MIT Press.

Brinken Dozier, J., Husted, B. W., McMahon J. T. (1998). Need for Approval in Low-context and High-context Cultures: A Communications Approach to Cross-cultural Ethics. *Teaching Business Ethics,* 1998/2.

Brown, M. T. (2006). Corporate Integrity and Public Interest: A Relational Approach to Business Ethics and Leadership. *Journal of Business Ethics*, 2006/66.

Buber, M. (1970). *I and Thou.* New York: Charles Scribner's Sons.

Deetz, S., Cohen, D. and Edley, P. P. (1997). Toward a Dialogic Ethic in the Context of International Business Organization. In F. L. Casmir (Ed.), *Ethics in Intercultural and International Communication.* Mahwah, New Jersey, London: Lawrence Erlbaum Associates, Publishers.

Floistad, G. (1999). Ethics in Visioning and Modelling by Leaders. In S. K. Chakraborty & S. R. Chatterjee (Eds), *Applied Ethics in Management. Towards New Perspectives.* Berlin, Heidelberg, New York: Springer Verlag.

Freeman, E. R. (1994). A Stakeholder Theory of the Modern Corporation. In E. R. Freeman, *Ethical Theory and Business.* London: Prentice Hall.

French, W. & Allbright, D. (1998). Resolving a Moral Conflict Through Discourse. *Journal of Business Ethics*, 1998/17.

Gardiner, M. (1996). Alterity and Ethics. A Dialogical Perspective. *Theory Culture & Society*, Vol. 13, No. 2, May 1996.

Hermans, H. J. M. & Kempen, H. J. G. (1993). *The Dialogical Self. Meaning as Movement.* San Diego, London: Academic Press, Inc.

Leach, E. (1976). *Culture and Communication: The Logic by Which Symbols Are Connected. An Introduction to the Use of Structuralist Analysis in Social Anthropology.* Cambridge: Cambridge University Press.

Leube, K. R. (Ed.) (1987). *The Essence of Friedman.* Stanford, CA: Hoover Institution Press, Stanford University.

Levinas, E. (1980). *Le Dialogue: Conscience de Soi et Proximité du Prochain.* Roma: Instituto di Studi filosofici.

Levinas, E. (1989). *The Levinas Reader.* Oxford: Basil Blackwell.

MacIntyre, A. (1981). *After Virtue. A Study in Moral Theory.* London: Duckwort.

Myerson, G. (1994). *Rhetoric, Reason and Society. Rationality as Dialogue.* London, Thousand Oaks, New Delhi: Sage Publications.

Pruzan, P. & Thyssen, O. (1990). Conflict and Consensus: Ethics as Shared Value Horizont in Strategic Planning. *Human Systems Management.* pp. 135–151.

Pruzan, P. (1999). From Control to Value-Based Management. In S. K. Chakraborty & S. R. Chatterjee (Eds), *Applied Ethics in Management. Towards New Perspectives.* Berlin, Heidelberg, New York: Springer Verlag.

Rapport, N. (1997). Edifying Anthropology. Culture as Conversation; Representation as Conversation. In J. Allison, J. Hockey, A. Dawson (Eds), *After Writing Culture.* London, New York: Routledge.

Sarbin, T. R. (1993). "Foreword" to Hubert J. M. Hermans' and Harry J. G. Kempens's *The Dialogic Self. Meaning as Movement.* San Diego, London: Academic Press, Inc.

Silverman, H. J. (Ed.) (1991). Gadamer and Hermeneutics. *Continental Philosophy IV, Science. Culture. Literature: Plato, Heidegger, Barthes, Ricoeur, Habermas, Derrida.* New York, London: Routledge.

Thomilson, T. D. (2006). Monologic and Dialogic Communication. *Dialogue* IPC 301. http://umdrive.memphis.edu/ggholson/public/Dialogue.html, 18 November 2006.

Eleanor O'Higgins
University College Dublin and
London School of Economics

Yvon Pesqueux
CNAM Paris

Management Education – A Path to Business Integrity?

The wave of business scandals in the early 2000s, such as Enron and Worldcom, followed by the implosion of the global financial system in 2008, has caused much soul-searching within the business community itself as well as among its stakeholders, including society-at-large. These crises of confidence in business have entailed dramatic examples of corporate greed and disregard for the exercise of any ethical principles.

Were the ethical failings we have witnessed attributable to a few rogue individuals, or were they systemic, based in common management values and practices? Probably, each explanation amplifies the other, whereby the business system and its underlying assumptions offers fertile territory for ambitious self-seeking individuals who can make their companies look good. Tightening up regulation appears to have been the remedy of choice in an attempt to preclude further malpractice in virtually all capital market jurisdictions.

This paper first examines how contemporary management philosophies give rise to the practices that have created scandals and business failures. The second section depicts the current role of management education in perpetuating the situation. Then the final section of the paper asks whether it is possible for management education to reverse its role to play a part in creating a more honourable business environment.

1. Current Management Philosophies and Practices

The driving forces behind contemporary management philosophies and practices are:

- 'managerialism' and 'performativity';
- economics underpinning;
- powerful stakeholders;
- the nature and dissemination of management literature.

1.1 'Managerialism' and 'Performativity'

Generally speaking, business has become a predominant force in society, not only in traditional capitalist countries, but also (and possibly even more prominently) in formerly anti-capitalist ones, like China and Russia. Coincidently, since the 1970s, we have witnessed an extraordinary expansion of management sciences. Increasing numbers of students attend management education programmes, bestowing a growing significance on management science. Implicit in the dominance of business and its underlying management science is the notion of 'managerialism', the ideology through which everything is organized according to managerial categories (e.g., New Public Management). This is based on knowledge of the patterns or laws that govern business, a kind of 'causal determinism' (Ghoshal 2005, p. 77), that regards people as behaving mechanistically. This suggests that management is at the core of the ability to control events, and create the performance that characterizes the success of business. The success sought is on behalf of corporations and their managers, rather than on behalf of a broader array of stakeholders or society itself (Grey 2004).

1.2 The Role of Economics

In effect, the implicit models adopted by managers have taken the human being and human will out of business behaviour. Despite evidence of the limitations of Homo Oeconomicus to explain the totality of behaviour, this model of the dominance of self-interest as governing all transactions persists (Ghoshal 2005, Zsolnai 2002). Management theories and transactions are overwhelmingly geared toward controlling behaviour in the name of shareholder value, efficiency, etc. Thus, the theories-in-use are agency theory, transaction cost theory, and various game theories where business is seen as a zero-sum game (Ghoshal 2005). Although business is basically a human activity, it is not recognized as such with the objectification of the person, for example, invoking terms like 'human resources' or 'human capital' in management jargon.

Even if this illusion of control were reality, it does not speak of its ultimate purpose. Does managerialism have a purpose beyond 'performativity' (Fournier and Grey 2000), and/or the self aggrandizement of managers? Is there a social purpose beyond shareholder value? Various observers comment on the moral muteness of contemporary management theories (Ghoshal 2005; Anonymous, 2004, quoting Henry Mintzberg). Indeed, the nature of management thinking encourages unethical practices where the ends of 'winning' justify the means (O'Higgins 2004). In fact, the morality of the means is never really considered.

It is instructive to consider the rankings of the world's most admired or respected companies, in *Fortune* magazine and *The Financial Times*, respectively, as ranked by CEOs, market investors and non-governmental organizations (NGOs). All of these observers tend to rank companies like Microsoft and General Electric and their leaders in the top positions, simply by dint of their renown as being highly profitable and creating shareholder value. The fact that Microsoft's business tactics have been questioned on both a legal and ethical basis does not seem to matter.

While contemporary management thinking does purport to rely on an array of social sciences for its knowledge, in fact, its most basic foundations are built on economics. This views the power of markets, based on self-interest as the most desirable foundation on which to model business behaviour. This kind of advocated behaviour becomes self-fulfilling because it creates the kind of conditions that cause it to be true (Ferarro, Pfeffer, and Sutton 2005). An example is the widespread acceptance of agency theory which gave rise to the use of stock options, designed to align the interests of managers with shareholders, on the assumption that such external incentives are required to curtail managers taking advantage of their positions. To an extent, it has been shown that self-interested behaviour is learned, for example, by taking courses in microeconomics! Once such self-fulfilling behaviour is established in a self-reinforcing vicious cycle, it becomes almost impossible to disprove and replace.

The genesis of the disciplinary field of management rests in a grey zone between management education, managerial research and consultancy. The constitution of this grey zone is closely knit with the nature of a management focus over practices as well as to the modalities of its funding. Through considerable funds mobilized by executives and federations of companies, well endowed sponsors have a major influence on the construction of the body of management knowledge, leading to 'corporatization' of all kinds of organizations. An example is the almost universal admiration for Enron and its management in the heyday of the company, not only by markets and managers worldwide, but also by business gurus and management educators alike. Those who dared to question were quickly and ruthlessly dismissed by the stampede of Enron fans.

The reaction to scandals has been to institute rules and regulations built on the premise of self-interested individualism, instead of a rounder more comprehensive appreciation of human behaviour that includes a moral sensibility. Even when social sciences other than economics are introduced into management, they are subordinated to the 'economic man' perspective, since the legitimacy given to utilitarian reason and its related 'techniques' has the effect of downgrading the 'humanities'. Management science is elaborated by borrowing concepts and methods from other fields, often impoverishing them of their original understanding without any scruple, before reinterpreting them to arrive at management applications and techniques. An example is the concept of culture. Managerial science has attributed a 'culture' to organizations and took this concept from anthropology, without considering its history or its categories. The anthropologists were then forced into incorporating organizations as a research field, progressively adapting their concepts and even their discipline to the categories of managerial science. Thus, 'culture' is applied indiscriminately, especially in the management literature on organizational change, whereby people (human resources) become objects whom management can shape as befits the economic purposes of the organization. This example illustrates a kind of 'passive' domination of social sciences by management science.

Interestingly, the traditional managerialist model is now being presented with a new element – Corporate Social Responsibility, now elevated to the point where it has its very own acronym, CSR. Another popular term is Corporate Citizenship. Basically, those who espouse CSR are reacting to the reputation damage suffered by business enterprises in the wake of the various high profile scandals. Other pressures to behave responsibly come from stakeholders, activist NGOs and general societal and institutional pressures (Waddock, Bodwell, and Graves 2002). Many companies have climbed on the CSR bandwagon, issuing CSR and Stakeholder reports. Simultaneously, the CSR measurement/assessment/accountability industry has burgeoned. The emphasis here is *communicating* that the company is responsible rather than actually *being* responsible. At worst, CSR is nothing other than a façade for management control (O'Dwyer 2003).

It is intriguing to witness the rationale offered by companies for becoming 'responsible'. Often, the 'business case' is presented, i.e., enlightened self-interest, for defensive purposes, or for enrichment purposes, to win over customers, enhance reputation, increase efficiency or innovation, etc. (e.g., Cowe and Hopkins 2003). One rarely hears the case that to act responsibly is desirable in its own right and should be a fundamental principle of business, whether or not there is a 'business case'. Thus, paradoxically, the CSR movement has the effect of reinforcing managerialism rather than alleviating it.

The model of management advocated by the prevailing wisdom fails to differentiate between contexts, and takes for granted that all management occurs within the US system (Starkey and Tempest 2005). How has this domination of managerialism, with its roots in American conservative capitalism, managed to insinuate itself into societies on every continent of the globe? Firstly, we have globalization itself as a vehicle for spreading the philosophy through its associated practices. In newly capitalizing countries, initial exposure to these practices offers the impression that this is the only way to behave, especially if local companies wish to emulate the success of their Western counterparts.

1.3 Management Literature

Another mode of transmitting managerialism has been through the management literature. The development of a managerial literature signals an expansion of management science and also demonstrates a social practice. One sign of this is the creation of a new 'management' category in classrooms, libraries, bookstores, and newsagents. The literature is one of the aspects of the transition from managerial practices to a 'managerial knowledge'. The emergence of management science gives a concrete expression to the attempt to build a unified teleological optimistic model of a successful organization. Does managerial knowledge originate with the scientific community or management practitioners? Until recently, it was accepted that it was induced by practitioners. Today, it is not so clear. Managerial science has become the place of an understanding, a reinterpretation of practices and also the place for modelling new practices. Their transcription in managerial literature(s) has the effect of embedding them as 'truth'.

If practices establish a body of knowledge, built on what is perceived as effective techniques, the passage from the knowledge of practitioners to the knowledge of scientists and, by extension, to the knowledge of a science, is strongly dependent on a social game in which the literatures play an essential role. The body of knowledge may qualify a profession, and often ends up with academics as 'the knower'. However, the academic is also the figure which qualifies/decrees 'science' and what is 'scientific'. This is how management science constitutes itself between practices (know-how) and a science governed by academics. The managerial literature also raises the question of the distortion associated with the mediation of a medium, as the passage to the written form fixes the observed phenomena in a durable way through identification and objectification.

A classification of the managerial literature can facilitate our understanding of its role in perpetuating a particular view of business and management.

'Best sellers', airport shops books. This category contains a diversity of material. One is biographies/autobiographies/leaders' testimonies – always leaders. The promises made within this literature often surpass their delivery. This literature possesses the status of a testimony – stories are taken as certitudes, with the author/hero as the key character of the stage. He is never supposed to have re-qualified his actions. The stories are often accompanied by non controversial and non debatable prescriptions, because it is leaders who assert them. These books have the status of exemplarity. They offer 'magic thoughts' where 'memories' serve as 'compost' for the elaboration of an ideology as a legitimization of its assertions. Examples are the books written by Jack Welch the former CEO of General Electric.[1]

Other types in this category are bestseller 'recipe books'.[2] Consultants' predictions and prescriptions are found within this literature. This literature has status and recognition, even if much of it is ephemeral. These books contribute to the construction of managerial fads and fashions. They can be also be interpreted as a first degree of conceptualization in a project of passage from practices to concepts with more scientific validity. Their self-predictive capacity is important because managers tend to act according to these prescriptions. Authors usually use their books as an advertising tool for their consulting and training activities.

Books written by academics. This category includes textbooks that are used for business degree and diploma courses. Their contents evolve in relation to their fields, incorporating the 'personal touch' of their authors. They offer a summary of the conceptualizations in the field. There are also specialized books which give some view on a specific question or a new vision of the aspect they target. It is one of the ways of institutionalizing and expressing a criticism.[3] Other books offer syntheses on the history and the development of a question or on the links established between foundations and current developments (e.g., handbooks or encyclopaedias). Positions are often elaborated by authors to validate their status as reference. This allows the authors to distinguish themselves from others in the same area. This aspect of managerial literatures contributes to the 'scientism' of management science.[4]

Practitioner periodicals, journals, magazines and newspapers. Here we find accessible management periodicals, such as *The Harvard Business Review*; these may offer the first conceptualization made from practices, using illustrations of anecdotal evidence and case studies. This first conceptualization is mostly prescriptive, and may be deterministic. This literature drifts between understanding and proposing, between theories and prescriptions. Other popular business media, for example, specialist financial newspapers such as *The Financial Times* or *The Wall Street Journal,* emphasize company and markets current events. General newspapers also feature a business and companies section. They all contribute to the solidification of practices into knowledge. Other periodicals give a more detailed analysis on selected socioeconomic, political and business events and on existing practices, iconography included, for example, *Business Week, The Economist.* The centrality given to company events and the large number of titles and readers have made this press important. In turn, this press privileges particular opinions and interests, since it is a medium for advertising from these interests, directly and indirectly. Thus, it is an important vector for the production of an ideology. It is radically taken as telling the 'truth', reinforced by emotional logic.

1.4 Academic Journals

Here, the project of 'scientificity' is very clear, determined by the social conventions of its academic environment. The main objectives of the game are to be published and cited by others. By-and-large, the content of these journals is written by academics for academics in a very stereotyped style. Paradoxically, it is divorced from the practices it purports to analyze and is of low interest to practitioners. This is a virtually unbreachable gulf (Bartunek 2007). Ghoshal (2005) deplores the application of analytic methods from the physical sciences to the social sciences in management research, subjecting everything to the causal mode, and excluding intentionality, a necessary condition for moral thinking and choice. In fact, the body of knowledge thereby produced has been deemed 'the pretence of knowledge', devoid of human mental phenomena and, indeed, of common sense.

Again, as a case-in-point, agency theory has dominated thinking in academic research despite its 'unrealistic assumptions and invalid prescriptions' (Ghoshal 2005, p. 81). This is because alternative theories, such as stewardship theory are too complex to yield to mathematical modelling and its comrade, reductionism. These latter elements are de rigueur in 'leading' academic management journals, whose 'impact' is assessed by the number of citations in the same genre of journals, at the expense of the scholarship of common sense. John Kay (2006) and Jeffrey Pfeffer (2007) illustrate how peer review becomes inward looking, and how 'a further step down a well defined road wins easier acceptance than a deviation from the beaten track'. Hence, paradigm shifts, and any form of innovative thinking are discouraged and what passes for excellence in publications is often nothing more than self-referring formulaic accepted wisdom, with some meaningless new increment.

2. The Role of Management Education

By-and-large, management education as presented in the medium of business schools globally has acted as a cheerleader for the management models and practices portrayed above. This is manifested and reinforced by:

- the managerialist perspective;
- deference to powerful stakeholders, especially companies/prospective employers of students, and students themselves as 'customers' or 'consumers';
- simplistic tools and formulas to train students
- academic tenure and promotion systems;
- the extremes of popularist science and pedantic science.

The MBA programme is usually the flagship degree for university based business schools. It is the vehicle through which current managerialism is perpetuated, since students take MBAs in the expectation that it will position them to get a well-paid job, on the assumption that an MBA degree equips its holder to be an effective manager, i.e., to create value in the company by being profitable and impressing shareholders. Competitive advertisements for MBA programmes promote this expectation. For example, 'We deliver a one-year MBA which provides a real return on your investment in time and money' (Cranfield); 'More Business Acumen' (Henley); 'reward is the clarity to triumph, whatever moment of truth you face' (Chicago); 'We add value to your future. We add a future to your value.' (Bocconi).[5]

Generally, business schools divide management subject matter into applied specialties, such as finance, marketing, management information systems, etc. What is taught is a managerialist perspective on using these specialties to enhance performance. Finance is a good case-in-point. Finance enjoys a central and prestigious position in the business school curriculum, since many MBA graduates thirst after jobs in prominent investment banks. 'And finance is seductive. It is numbers intensive, based on fundamental mathematical principles, and seems values neutral, almost scientific. Yet finance involves substantial moral judgements about the world in which we live. And financial tools can help to solve deep social and environmental problems' (Aspen Institute 2006). One can easily explain the subprime crisis engendered by reckless lending and debt repackaging practices, as arising because bankers, devoid of moral considerations, used these techniques for short term gain and rewards. Other practices, such as short-selling and the use of financial instruments, further and further removed from the productive economy, are similarly explained. Now, various stakeholders deplore the fact that those who enjoyed the exorbitant rewards from irresponsible behaviour, are bailed out by taxpayers. In some cases, the CEOs who stood over this behaviour have walked away with huge bonuses. A case-in-point is Stan O'Neal who left behind writedowns of $14.1 billion as he walked away with a compensation package valued at $161 million.

Decrying the lack of 'professionalism' in business education, Trank and Rynes (2003) describe how business schools are captive to the 'unprofessional' models extant in the business world today. This is because of pressures on business schools from various stakeholders. Since they provide the financing to business schools, either directly or indirectly, future employers and management students seeking employment are two extremely powerful stakeholder groups which impose their agendas on business schools. Overwhelmingly important are the demands from business for immediately useful skills and techniques from graduates, as opposed to an appreciation of the complexity of business as a human activity. On their part, business students enhance their employability by showing that they have mastered skills relevant to employers (i.e., simplistic techniques) and have received brilliant grades, to look good to the best employers. In an exceedingly competitive environment for business schools, management educators are encouraged to pander to the desire for simplistic techniques and high marks. Many business schools employ adjunct faculty who are practitioner executives or consultants to share their experiences and 'war stories' with students. Usually, these practitioners are exemplars of success in the Darwinian business world. While very popular and entertaining, one must question what sustainable learning comes from these learning episodes, which are basically at the level of 'pub talk'. It follows that any coursework which attempts to question current management philosophies and practices or to provoke thinking about human intentionality and morality is discouraged as a distraction from the essential goals of business education. It would be a minority taste, at best.

Conflicting demands on business school faculty also play a part in perpetuating the ethical void. On the one hand, business school academics have to be 'relevant' to students seeking employment. Consequently, evaluations of teachers by students are usually based on a reputation for giving high grades and entertainment value in the classroom. So, there is pressure to cater for the lowest common denominator.

On the other hand, academic tenure and promotion is based on publications in leading 'scholarly' journals. The pecking order of journals dictates that those which are 'practitioner oriented' do not carry as much academic weight as those which are more 'conceptual' and/or based on positivist research with large samples. Each article is expected to include a 'theoretical' contribution, no matter how trivial or divorced from reality (Hambrick, 2007). However, to achieve the large samples so esteemed by these journals with primary data is too laborious and time consuming. The difficulty is underlined when researchers require the co-operation of management to discover at some depth what is going on. At the same time, research must be objective, examining warts and all, 'the establishment of a more reflexive, critical approach to research' (Starkey and Tempest 2005, p. 74) – a difficult path to navigate. Skills in working and negotiating with practitioners on a meaningful research agenda without compromising scholarly integrity does not appear to be part of the curriculum for Ph.D. students in management. Instead, there are ever more sophisticated courses on the 'scientific method' as if studies in management are identical to those in the physical sciences, a false premise according to Ghoshal (2005). Since it is easier to gather information from vast databases than from engagement with human practitioners, much published research is based on such impersonal material. What is taken as conclusive may be statistically correct, but substantially pointless, since it treats the human actor only as a black box, whose desires, emotions and actions are to be inferred from a statistic based on regression analyses or mathematical models. The findings from this type of research tend to be hollow and irrelevant, essentially 'pedantic science' (Hodgkinson, Herriot, and Anderson 2001). Even to the practitioner to whom it is intellectually accessible, these types of publications tend to turn practicing managers away from academics, whose 'street cred' is thereby reduced. Such lack of credibility is bound to extend to any moral sense that management educators might wish to impart to students.

Thus, we have the predicament of business school faculty members, on the one hand, having to utilize 'populist science' literature (Hodgkinson,

Herriot and Anderson, 2001) to get through to their students, but eschewing this literature to distribute their own work if they wish to gain acceptance among their peers. Trank and Rynes (2003) propose that teachers should share their research knowledge with their students and ground their students in research methodology that would make them question the oversimplified notions found in popular management literature. While the latter objective is surely a worthy one, the nature of most current management research and the way it is presented is hardly an alternative for imparting meaningful insights to students, especially those with ambitions to be effective practitioners. Trank and Rynes are exactly right in advocating deeper understanding of management and organizational life, but unfortunately, much academic literature is either too simplistic or too irrelevant to impart this type of understanding (Starkey and Tempest 2005).

Neither of the conflicting pressures on academics in business schools – the demand of relevance/popularity with students versus the obligation to publish in particular journals – carries much interest in the human moral condition, for example, what people in organizations with a difficult choice to make, go through, much less that these people have a personal history and that this interacts with their current life situation, of which their working life is a part.

However, the business scandals of the early 2000s have awakened business schools to their failure to provide any moral compass to future managers, as they taught the skills required for the successful economic man, to the exclusion of the human person. The reaction has been to introduce courses in business ethics and corporate social responsibility. While these are commendable, the motives behind these courses are often for public relations purposes, a box ticking exercise whereby the institution involved is able to state that such a course is offered when the question is posed. Such courses may not be supported as avidly as those in other tools and techniques oriented subjects and may be divorced from these other subjects in the curriculum, whose material fails to integrate any sense of ethics or social values.

3. Management Education and Business Integrity

Given the current positioning of mainstream business schools vis-à-vis the inculcation of integrity into business, is it realistic to consider that management education can actually make a contribution to enhancing business integrity? Despite the obstacles, we suggest that certain levers are available to enhance business integrity through management education:

- business-education links with practitioners and other social actors;
- universities as venues for critical thinking;
- universities as providing grounding in social sciences additional to economics;
- professionalized knowledge as 'pragmatic science', professionalized values and ethics;
- business school based ethical awareness courses and modules;
- business school based career services which facilitate informed, personally relevant career choices.

Firstly, we can consider that management education does not have to be university business school based. However, Starkey and Tempest (2005) suggest that the university is the best venue for fulfilling three requirements for management education – skills of analytical analysis, interpersonal skills and a body of knowledge about organizations. This should produce a critical reflexive practitioner with a wide in-depth understanding of social, political and ethical dimensions of organizations. The achievement of such an understanding requires an inclusive arena comprised of stakeholders from industry, politics science, society and various other community groups. It is suggested that the university is an ideal venue for bringing together these stakeholders to exchange information. Thus, faculty members become knowledge development and communication facilitators among stakeholders. By joining into this disciplined dialogue, managers begin to understand what is a good life and its associated values, for example, trust.

However, the university based business school cannot achieve this status unless it is willing to abandon unthinking loyalty to managerialism and its implications for management practices and its relationship with society. The business school must open itself to a broader array of interests and perspectives. It means that management has to be embedded in society and daily life, not merely considering society as an 'environment' to be operated upon.

The integrity of business may be improved by other means whereby the education process may play a part. It has been suggested that the 'professionalization' of management will imbue it with ethical standards (Khurana, Nohria, and Penrice 2005; Trank and Rynes 2003). However, when one examines the hallmarks of what constitutes a profession, it is difficult to see how management can qualify. A common body of knowledge resting on a widely accepted theoretical base that can be applied in various contexts and situations is one of the hallmarks of a profession. This knowledge is usually imparted in university faculties, for example, medicine or law. However, it has been argued that management has no underlying body of knowledge or accepted theory. Future employers suggest that management education should be practice based rather than knowledge based. The MBA degree, while advantageous in procuring a desirable job in many countries does not offer a fundamental body of knowledge, without which management cannot be practiced. A related characteristic of professions is a certification system to exclude the unqualified, i.e., those without the requisite knowledge as proven by passing some kind of test or university degree. This is another 'professional' convention that would be impossible to enforce in the management sphere.

Other distinguishing features of professions revolve around values and ethics. One of these has to do with commitment to public service beyond self-interest. Khurana et al. (2005) show how this requirement is violated by the acceptance of the agency model which promotes self interest, and highlighted by the business leaders who have enriched themselves by means of unfair and distortionary pay packages. Finally, professions are normally governed by an enforceable code of ethics. Management is too amorphous an occupation on which to impose a uniform code of ethics. We have seen that corporate codes have had little impact on shaping ethical behaviour by the likes of Enron, which had an elaborate code. Hence, while certain aspects of professionalization of management are very appealing as a means of invoking a moral sense, we cannot transform management into a profession, since management does not, and is unlikely to be, able to meet the criteria that distinguish professions.

To dismiss management as a formal profession is not to ignore some aspects of professionalism that can be adopted by management science and the way it is taught and researched in university business schools. One such aspect relates to the body of knowledge. Trank and Rynes (2003) suggest that understanding management as opposed to touting fads and simplistic formulas is the way forward. However, the understanding will not come from the leading academic journals whose rigorous research is filled with structure but no substance of interest to practitioners. Instead, journals which speak to present and future practitioners should embrace 'pragmatic science' (Hodgkinson, Herriot, and Anderson 2001), knowledge that is simultaneously rigorous and relevant. This is consistent with Starkey and Tempest's (2005) strictures to academics to engage with practitioners on issues that mean something to those practitioners. Academics can help practitioners to make sense of their experience, thereby working with 'theories in practice'. However, Hodgkinson et al. (2001) point out that research training in academia today does not provide the interpersonal skills to engage with other stakeholders in primary research. This kind of skill is a prerequisite to conduct this research.

Part of this promotion of understanding should engage fundamental disciplines (in addition to economics), although it is not customary for management science to articulate with social sciences other than economics. Of course, the university is an eminently suitable place to assimilate all the social sciences. The other social sciences that may be invoked are sociology, political science, anthropology and psychology. Sociology is interested in the understanding of life in society, without prescriptive aim, with subsets, which require the construction of concepts and specific methods, for example, sociology of organizations and sociology of labour.

Anthropology tries to explain the modalities of functioning of a 'natural group', seen as the elementary shape of life within societies. These 'natural groups' have been found in 'primitive societies' and considered as a blueprint for social life in modern societies (see, for example, Redfields, Linton, and Herskovits 1936, Mead 1962, Bastide 1970, Geertz 1973, Leroi-Gourhan 1974, Evans-Pritchard 1986). Can the organization then be considered as an elementary group or a set of elementary groups? It is from this perspective that, for example, the idea comes that organizations have cultures, rites etc. To paraphrase Douglas (1986), understanding organizations is to understand how organizations think, i.e. how they live with culturally rooted, apparently unquestioned norms. Her famous example of the tomato is that tomatoes have been classified as vegetable. Had they been classified as a fruit, their destiny would have been completely different.

Psychology is at the root of human emotions and behaviour. Invoking psychology beyond the simplistic economic man dimension normally presented in business schools as organizational behaviour would be helpful to understanding business in its complex human manifestation, for example, the foundation and practice of leadership. Management science also suffers from 'pop psychology', which has insinuated itself into areas such as the manipulation of groups/teams and the formulaic misuse of psychometric testing, without really appreciating the psychological constructs in all too often computer generated personality profiles. Similarly, we would wish to get beneath the stimulus-response school of psychology, at the heart of the techniques and tools approach to management.

Once one considers the dynamics of the relationships among individuals, subgroups, organizations and society, the discipline of political science is also pertinent, as it refers to the will of the body politic – an element that is increasingly demanding higher standards from the business community and the integration of nonmaterialistic values (Giacalone 2004).

The university setting is also one that can lend itself to the examination of alternative paradigms, through critical inquiry. This questioning approach might be applied to the domination of managerialism, letting in a broader view of the role of business in society and interest in the welfare of a wide range of constituencies/stakeholders. This view would connect another facet of professionalism, the notion of the good of the community and how business can be of service to society as a primary aim, not as a means of self promotion, as is currently the case with much CSR activity.

The recommendations for pragmatic science and the invocation of alternative paradigms are not easily implemented, since vested interests have the effect of perpetuating both pedantic science and the current managerialist model. It is up to leaders in business schools to encourage and reward pragmatic science and new paradigms as published in alternative journals and media to those which insist on the status quo.

Business education can also be instrumental in helping practitioners to grasp and resolve ethical issues in management. Robert Giacalone (2004) suggests a transcendent business education model to go beyond profitability and self-interest. Such an education is founded on five goals – empathic understanding, generation of a better world, mutuality of effort and gain sharing, civil aspiration that surpasses mere compliance or avoidance of wrongdoing, and intolerance of ineffective humanity, i.e., of people who fail to respect the humanity of others. Hartman (2006) advocates a concentration on developing character or virtue ethics through business ethics courses. Such courses would concentrate on coherence between values and life interests to create a state of well-being. The pedagogy by which this might be achieved is 'reflective equilibrium' by the judicious use of case studies. People develop their ethical reasoning abilities and sensibilities through exposure to ethical issues, especially through personal experience (Trevino and Nelson 2007). Therefore, discussion of real life events and dilemmas experienced by students can inform their ethical thinking and overall consciousness of moral life. This sort of pedagogy should permeate conventional business courses to create an integral grasp of ethical themes in business. Additionally, business school teachers can promote examples of positive role models, leaders and companies who embody the five goals of transcendence. This is in contrast to the current lists of admired/respected companies, which tend to concentrate around financial success, to the exclusion of broader societal considerations.

A duty often undertaken by business schools is the development of the person, especially in MBA education. This sort of development often involves skills based, ungraded courses, such as presentation skills. Ethics courses could fit well into the personal development mode. Business schools also provide career services for their students, to help them find employment. In fact, it has been found that students prioritize quality of career service in evaluating their MBA programmes. The Aspen Institute (2006a) has called attention to career services in business schools and the extent to which they influence the meaning of work to management graduates. They are helpful at placing graduates into jobs in nonprofit and public sectors and traditional CSR disciplines, such as public affairs and corporate foundations, and at finding partners to provide work experience and internships in those kinds of organizations. However, they are less involved in helping students find opportunities in mainstream business careers that will create social and environmental value alongside the financial bottom line. Nor do these services tend to adopt a critical view of the ethical standing of potential employers, or to incorporate the social and ethical aspirations of the school itself into their advisory role to students. Hartman (2006) shows that part of education for virtue is for students to choose workplaces that do not undermine their character. Thus, career services, as a necessary adjunct to management education, have the potential, as yet far from fully realized, to attune future managers to ethical and social issues and values.

4. Conclusion

We have seen how managerialism has created a void where ethical contents should be, and how management education has perpetuated this situation, either actively by mimicking the managerialist model, or passively, by retreating into pedantic science. Paradoxically, the 2008 breakdown of the global financial system offers an opportunity for a fresh start in management thinking and education, as the moral bankruptcy of the previous system has been recognized. The situation will be difficult to reverse, but a transformation in the underpinning and practices of management education will enable it to make a contribution to business integrity. Some of our suggestions centre around new links between the business school and society, between research and practice and between the business school and business itself (Starkey and Tempest 2005). We call for changes in internal processes and practices in business schools that will enable the new links to flourish. This should animate future business practitioners who prioritize integrity between ethical values and the lives they lead within the business world and beyond.

Notes

1 For example, *Winning* (2006), *Jack: Straight from the Gut* (2001), *Jack Welch Speaks: Wisdom from the World's Greatest Business Leader* (1998).

2 For example, *The One Minute Manager* (1981) by Blanchard & Johnson, updated as *High Five* in 2001).

3 Examples are the books by Charles Hampden-Turner & Fons Trompenaars, *Managing People across Cultures* (2004) *and Riding the Waves of Culture* (1997).

4 A prominent example is the influential Michael Porter depicting his model of competitive advantage through his books, *Competitive Strategy* (1981) and *Competitive Advantage* (1985). These books have assumed the status of 'classics'.

5 Advertisements in *The Economist*, 2006. This illustration shows the systemic link between all the parties that support managerialism, including the media.

References

Anonymous (2004). Mintzberg's Moral Maze. *Director,* September, pp. 84–86.

Aspen Institute (2006). *A Closer Look at Business Education: Finance.* Business and Society Program, February.

Aspen Institute (2006a). *A Closer Look at Business Education: Career Services.* Business and Society Program, February.

Bartunek, J. M. (2007). Academic-practitioner Collaboration Need Not Require Joint or Relevant Research: Toward a Relational Scholarship of Integration. *Academy of Management Journal,* 50(6), pp. 1323–1333.

Bastide, R. (1970). *Le prochain et le lointain.* Paris: Cujas.

Cowe, R. & Hopkins, M. (2003). *CSR: Is there a business case?* London: ACCA UK.

Douglas, M. (1986). *How Institutions Think?* New York: Syracuse University Press.

Evans-Pritchard, E. E. (1986). *Les nuirs.* Paris: Gallimard.

Ferraro, F., Pfeffer, J., Sutton, R. I. (2005). Economics Language and Assumptions: How Theories Can Become Self-fulfilling. *Academy of Management Review,* 30(1), pp. 8–24.

Fournier, V. & Grey, C. (2000). At the Critical Moment: Conditions and Prospects for Critical Management Studies. *Human Relations,* 53(1), pp. 7–32.

Geertz, C. (1973). *The Interpretations of Cultures – Selected Essays by Clifford Geertz.* New York: Basic Books Inc.

Ghoshal, S. (2005). Good Management Theories Are Destroying Good Management Practices. *Academy of Management Learning and Education,* 4(1), pp. 75–91.

Giacalone, R. A. (2004). A Transcendent Business Education for the 21st Century. *Academy of Management Learning & Education,* 3(4), pp. 415–420.

Grey, C. (2004). Reinventing Business Schools: The Contribution of Critical Management Education. *Academy of Management Learning & Education,* 3(2), pp. 178–186.

Hambrick, D. C. (2007). The Field of Management's Devotion to Theory: Too Much of a Good Thing? *Academy of Management Journal,* 50(6), pp. 1346–1352.

Hartman, E. M. (2006). Can We Teach Character? An Aristotelian Answer. *Academy of Management Learning & Education,* 5(1), pp. 68–81.

Hodgkinson, G. P., Herriot, P., Anderson, N. (2001). Realigning the Stakeholders in Management research: Lessons from Industrial, Work and Organizational Psychology. *British Journal of Management,* 12, Special Issue, S41–S48.

Kay, J. (2006). Conflicting Opinion Is What Drives Scientific Advance. *The Financial Times,* June 9, 15.

Khurana, R., Nohria, N. and Penrice, D. (2005). Is Business Management a Profession? *HBS Working Knowledge,* February 21.

Leroi-Gourhan, A. (1974). *L'homme et la matière.* Paris: Albin Michel.

Mead, M. (1962). *Société, tradition et techniques,* Paris: UNESCO.

O'Dwyer, B. (2003). Conceptions of Corporate Social Responsibility: The Nature of Managerial Capture.*Accounting, Auditing & Accountability Journal,* 16(4), pp. 523–557.

O'Higgins, E. (2004). Do Principles Count? In S. Crainer & D. Dearlove (Eds), *Financial Times Handbook of Management* (third edition, pp. 659–666). Harlow, Essex, UK: Pearson Education.

Pfeffer, J. (2007). A Modest Proposal: How We Might Change the Process and Product of Managerial Research. *Academy of Management Journal,* 50(6), pp. 1334–1345.

Redfields, R., Linton, R., Herskovits, M. J. (1936). Memorandom on the Study of Acculturation. *American Anthropologist,* Vol. 38, No. 1, pp. 149–152.

Starkey, K. & Tempest, S. (2005). The Future of the Business School; Knowledge Challenges and Opportunities. *Human Relations,* 58(1), pp. 61–82.

Trank, C. Q. & Rynes, S. L. (2003). Who Moved Our Cheese? Reclaiming Professionalism in Business Education. *Academy of Management Learning & Education,* 2(2), pp. 189–205.

Trevino, L. K. & Nelson, K. A. (2007). *Managing Business Ethics: Straight Talk About How to Do It Right* (fourth edition). Hoboken, NJ: Wiley.

Waddock, S. A., Bodwell, C., Graves, S. B. (2002). Responsibility: The New Business Imperative. *Academy of Management Executive,* 16(2), pp. 122–147.

Zsolnai, L. (2002). The Moral Economic Man. In L. Zsolnai (Ed.), *Ethics in the Economy*(pp. 39–58). Bern: Peter Lang.

Robert Elliott Allinson
Soka University of America

Value Creation as the Foundation of Economics

1. The Two Faces of Adam Smith

The argument of this paper, written by an ethicist and a philosopher, is that self-interest economics is fundamentally flawed and needs to be replaced by a spiritual economics or a value based economics[1]. Its argument contains two interwoven threads. One thread is an attempt to show why the fundamental philosophical notions of Adam Smith, taken as an illustration of self-interest economics, cannot lead to an equitable society.[2] Smith's *Wealth of Nations*, according to Jacob Viner, '…became a significant factor in determining the course of national policy not only in Britain but in other countries as well. This is much more than any other economic work has ever achieved; and Smith probably has had much more influence than any other economist.'[3] One wonders if it is Smith that Keynes had in mind when he famously quipped that all of us are slaves of some defunct economist. This despite Schumpeter's trumpeted dictum that 'the *Wealth of Nations* does not contain a single *analytic* idea, principle or method that was entirely new in 1776.'[4]

Whether single ideas or principles were new or not, the entirety of ideas that make up *The Wealth of Nations* was certainly new. And much turns on the meaning of the adjective 'analytic'. Was the 'invisible hand' an analytic idea? Was the notion that private interest adds up to public virtue (that self-interest on the part of the individual added up to the good of the whole) an analytic idea? If one takes Viner's definition of an analytic idea to be an idea that is rigorous, possesses internal consistency and bears a close analogy to abstract mathematical operations, then neither the invisible hand nor the private interest public virtue idea qualify as analytic ideas. Nevertheless they are extremely influential ideas. In this respect Schumpeter's dictum would appear to be of only minor importance.[5]

In his book, *Adam Smith's Moral Philosophy*, published with Cambridge University Press in 2005, Jerry Evensky gives an historical account of what he refers to the "Adam Smith problem" and categorizes Smithian interpreters into two main types. Amusingly, he refers to one as the Kirkaldian Smith (after his birthplace) and the other as the Chicago Smith (after the Chicago economists).

The Smithian version presented herein possesses more in common with the interpretation of such figures as the Nobel laureates' George Stigler and Jacob Viner. But it is not that they are Nobels that makes their interpretations compelling to me. Rather it is that the arguments that they co-advance, that without the emphasis on self-interest on the one hand and the invisible hand on the other of Smith's theory, that Smith's entire economic theory would collapse. For if self-interest as Stigler argues is the granite of the *Wealth of Nations*, then the invisible hand, as Viner implies, is the mortar. One recalls Stigler's famous sentence that begins his article, 'Smith's Travels of the Ship of State,' "The *Wealth of Nations* is a stupendous palace erected upon the granite of self-interest.'[6]

Evensky classifies the portrait of Smith that I present herein as the Chicago Smith, a portrait which he would say is painted by Frank Knight, Theodore Schultz, George Stigler, Milton Friedman, and Gary Becker. Strangely, Evensky does not mention Viner, whose stature among economists is monumental and whose interpretation of Smith is groundbreaking. In her introduction to her richly argued *Adam Smith and His*

Legacy for Modern Capitalism, Werhane refers to him as 'the well-known Smith scholar'. Robbins places him as 'probably the greatest authority of the age in the history of economic and social thought' and Blaug states that he was 'quite simply the greatest historian of economic thought that ever lived'.[7] Jacob Viner, who disavows allegiance with the Chicago School would have to belong to this category as well if we are to accept Evensky's scheme of categories. Evensky presents his interpretation of a more multi-dimensional Smith (my term) in which he aligns himself with Amartya Sen and James Buchanan (I would put Patricia Werhane here, as well).[8] But the multi-dimensionality aspect of Smith is not, as we shall see below, the driving force of his economic theory. It his self-interest aspect that is its driving force.

According to Glenn R. Morrow, there is no Adam Smith problem. For Morrow, Smith's ethics and economics work hand in hand if one remembers that prudence is one of the ethical virtues.[9] One could reply to Morrow that this merely moves the problem back into the ethics. More to the point, though, is that in order to achieve the economic success of the country, one must make sympathetic impulses subservient to the rule of egoistic ones. The full title of the volume, it is to be recalled, is *The Inquiry into the Nature and Causes of the Wealth of Nations*. The egoistic impulses must rule if economics is to be served.

The consequence of this, and therefore the real problem for Smith as seen by the present author is that one must treat economic relations as being between non-persons in order to submerge ethical impulses under economic ones. In so doing, not only does one *ex hypothesi* treat others unethically, it also, according to Aristotelian and Confucian notions, makes oneself unethical and helps to co-create an unethical society.

Some of the former point seems to be what Jacob Viner is addressing when he states that, 'According to Adam Smith the sentiments weaken progressively as one moves from one's immediate family to one's intimate friends, to one's neighbors in a small community, to fellow-citizens in a great city, to members in general of one's own country, to foreigners, to mankind taken in the large …'[10] It is ironic that it is the thought of an economist that can give rise to so many different and conflicting interpretations. One would think the postmodern hermeneutic theories of the possibility or rather the necessity of the infinite varieties of interpretation to which any text is susceptible applied more in economics than they did in their own field of literary criticism. In the field of literary criticism, the legacy of disciples has been far more slavish in the finiteness of their interpretations of their master's, Derrida's thought.

2. Equilibrium Revealed as the Invisible Hand

The second thread is a charcoal sketch of my theory of value based economics rather than profit based economics. It is offered in the spirit of shedding light on the foundations of a spiritual economics which I argue to be a value based economics and in the spirit of a guiding light to inform our discussions of business ethics.

It is clear that the vision of the *Homo Oeconomicus* is not proving to be a fruitful guide. Untrammeled greed, unsurprisingly, is not proving to be beneficial in an equitable sense. What happens to Smith's argument that it is good to follow one's self interest because it bests serves the interest of the whole if it turns out that the good of the whole is not served? Does that mean that one should not be driven by self-interest? For the post-Smithian capitalist, profit is to be maximized whether this serves the good of the whole or not.

Let us analyze the most fundamental idea of economics that according to the Nobel laureate economist George Stigler, 'is still the most important substantive proposition of in all of economics, that is the idea of equilibrium.'[11] The idea of equilibrium is roughly speaking the idea that all resources will tend to equalize over the long term. Supply and demand will reach an equilibrium with each other. There are two basic questions one can raise about this most fundamental law. Firstly, why should it be true? Clearly, it is a metaphysical law since most of the time (witness the current oil prices) resources are in disequilibrium rather than equilibrium. If there is a fundamental law, why should it not be rather that resources seek disequilibrium? Secondly, one can always ask, two questions, even if the law were considered valid: (i) At what cost in the short term and in the long term, and cost to whom, is this equilibrium reached? (ii) Who benefits from the preceding disequilibrium in the short term and its eventual equilibrium in the long term?

Equilibrium, as a notion, is value free. It does not signify the achievement of a better condition for some parties or for all parties. In addition, it is simply mechanical. It takes place due to the operation of market forces. It has no sympathies. It has no ethical preferences.

Equilibrium is, as one may have already begun to suspect, Adam Smith's robot hand.[12] As a robot hand, it has no leanings in any ethical direction. It is purely market driven. It distributes according to its iron law. It cares not that some receive an overabundance and others next to nothing. In the end, all will not receive a just share. In the end, there will be an equal supply of resources. But who will own these resources and who will not is left out of the mechanical equation.

The origin of the invisible hand theory may well be the idea of a Providential Order.[13] This might well be why the idea of an invisible hand did not occur to the ancient Greeks or the ancient Chinese. For the ancient Greeks and the ancient Chinese, the world was not ruled by Divine Providence. The invisible hand of Smith appears to be a secular version of Divine Providence. The problem is, whereas Divine Providence supposedly has human welfare at heart, can we ascribe such an ethical motivation to the robot hand?

According to Stigler, The idea of equilibrium is central to Adam Smith's theory. In Smith's own words, 'The market price of every particular commodity is regulated by the proportion between the quantity which is actually brought to the market, and the demand of those who are willing to pay the natural price of the commodity, or the whole value of the rent, labour and profit, which must be paid in order to bring it thither.'[14]

In Amartya Sen's more contemporary version, 'The market system works by putting a price on a commodity and the allocation between consumers is done by the intensities of the respective willingness to buy it at the prevailing price. When "equilibrium prices" emerge they balance demand with supply for each commodity.'[15]

But it is not a matter exclusively of demand in the case of Smith's version or of willingness in the case of Sen's. It is a matter of capability. Demand is not ethically driven. Demand is driven by the capability on the part of the purchaser of paying the price for the commodity. Supply is not ethically driven. The price of the supply is set by the supplier in accordance with what the market – read the financially advantaged – is capable and willing to pay and how high this price can be set by the supplier without losing sufficient volume of sales which would lower the profit margin. "Equilibrium prices" are at an equilibrium only for the financially advantaged.[16]

For the modern view of economics, distribution plays no part. According to Amartya Sen, this is true even of utilitarian welfare economics: In referring to his 1973 book, *On Economic Inequality*, he writes, 'Utilitarianism, which had been the mainstream approach to welfare economics, is profoundly unconcerned with inequalities precisely in the variable on which it focuses (and to which it attaches overwhelming importance) to wit, individual utilities. All that matters in the utilitarian view is the sum total of these utilities representing the respective individual advantages, independently of their distribution.'[17]

3. Adam Smith, the Moral Economist

For Smith, the general welfare is more a matter of faith than anything else. In his famous and often quoted phrase, the one motivated by his self-interest achieves more for the common good than the one who directly attempts to further the common good. There is no proof of this. It is an article of faith. What is interesting about this article of faith is that it endorses the idea of following one's self-interest because it brings about (or supposedly brings about) the good of all. Smith is not the champion of self-interest for self-interest's sake.[18] He is already a moral economist. His economics, and he is taken as the father of economics, is not profit for profit's sake. It is profit for the sake of the general good.

Let us quote, not the well-known passage regarding the self-interest of the butcher, the brewer and the baker from *The Wealth of Nations*, but instead, a lesser known passage from his moral tome, *The Theory of Moral Sentiments*: '[The rich] consume little more than the poor; and in spite of their natural selfishness and rapacity, though they mean only their own conveniency, though [because] the sole end which they propose from the labours of all the thousands whom they employ be the gratification of their own vain and insatiable desires, they divide with the poor the produce of all their improvements. [the trickle-down theory of the benefits of wealth, a more recent version of Equilibrium theory] They are led by an invisible hand to make nearly the same distribution of the necessaries of life which would have been made had the earth been divided into equal portions among all the inhabitants; and thus, without intending it [because they do not intend it], without knowing it, advance the interest of the society, and afford means to the multiplication of the species.'[19]

For Smith, self-interested economic behavior is inextricably linked to ethical outcomes. In particular, it is the rich, those who employ thousands for the sake of satisfying their own insatiable desires, who advance the interest of society. It is important to note that it is the division of society into economic classes that is the engine of the economic success of society. The drivers of this engine are the rich. And the fuel of the engine of the rich is insatiable greed. Let us review one of his statements from *The Wealth of Nations* in detail:'Every individual is continually exerting himself to find out the most advantageous employment for whatever *capital* he can command. It is his own advantage, indeed, and not that of society, which he has in view. But the study of his own advantage naturally, or rather necessarily leads him to prefer an employment which is most advantageous to society.'[20] (emphasis added)

Smith's core belief is that self-interest pursued for its own sake is necessarily linked to the good of society. It is this core belief that is to be questioned or at least qualified. For Smith, self-interest is linked to *material* advantage and it is this notion of self-interest that is to be challenged. For Smith, 'Every man's interest would prompt him to seek the advantageous, and to shun the disadvantageous employment.'[21] Smith's notion of advantage is not moral advantage. To make this point more clear, consider the first sentence of the previous quotation: 'Every individual is continually exerting himself to find out the most advantageous employment for whatever capital he can command.' He is not considering that capital should be put to use as philanthropy. For his next sentence is, 'It is his own advantage … which he has in view'. He is clearly referring to capital advantage and not to moral advantage. Ironically, it is the material interest of the individual that is taken to be the basis for an ethical outcome for the whole of society. The irrelevance of Smith's own ethical theory for his economics is put in a nutshell by Viner when he points out that, 'Nowhere in the *Wealth of Nations* does Smith place any reliance for the proper working of the economic order upon the operation of benevolence, the emphasis upon which was the novel feature in the account of human nature presented in the *Theory of Moral Sentiments*.'[22]

Smith, in his earlier *The Theory of Moral Sentiments*, also famously wrote, 'How selfish soever man may be supposed, there are evidently some principles in his nature, which interest him in the fortune of others, and render their happiness necessary to him, though he derives nothing from it, except the pleasure of seeing it.'[23] The problem, is, what happens when one's own advantage is in conflict with one's interest in the fortune of others? In this case, his latter book would seem to trump his former because it is, as he says, 'his own advantage [and by that he means material advantage] which is most advantageous to society'. But there is no need to pit one book against the other as is commonly done in the Smith literature. For as we recall it is stated in the former book that it is the rich, the greediness of the rich and the division into classes that define the economic success of society. And the doctrine of the invisible hand appears in the former book as well. It is important to emphasize that I am not chiefly concerned with Smith and his inconsistencies. There is already a literature devoted to this. Most of the literature consists of Smithian apologetics.[24] So much so that one thinks that 'The gentlemen do protest too much'. Why I use Smith at all is because he is known as the father of economics and because the fact that there is a tension between ethical impulses and materialistic self-advantages is illustrative of the fact that this is inherent in the very nature of a self-interest or profit based economics. It is not surprising that such a tension exists in Smith. It would be all the more surprising if it were absent.

4. Ancient Greek and Chinese Philosophy in Support of Spiritual Economics

At the very least, one must say, Smith is in conflict with himself. In the case of Plato, Aristotle and Confucius, no such conflict arises. In ancient Greek and ancient Chinese philosophy, ethics clearly takes precedence over profits. The idea that profit or advantage should be sought for its own sake and that by so doing society would necessarily benefit would be alien to classical Greek or classical Chinese thinking. For Confucius, when ethics and profits collide, one never chooses profit. It is as simple as that.[25]

In the *Symposium*, Plato described the human being as a lover, as a pursuer of beauty. For Plato, the ultimate experience one strove for in life was the experience of beauty and creation in the beautiful. To put this in another way, Plato saw the essence of the human being to lie in creation, in production, not in consumption. Plato thought that our ultimate experience, that for which life was worth living, was the aesthetic experience of the enjoyment and production of the beautiful. Plato realized that we were driven primarily by Eros. But Eros for Plato was not ultimately for material things. The highest stages of Eros were for the Beautiful and its products.[26]

Is it possible to say that great artists and thinkers, that Pericles, Michelangelo, Leonardo, Mozart, Van Gogh, Descartes, Spinoza and Marx were motivated by the urge to create in the beautiful rather than by the profit motive? Descartes died of pneumonia tutoring the Queen of Sweden at 5 a.m. in a cold Swedish winter. Spinoza, Mozart, Marx and Van Gogh died in poverty. If they had made profits, they would have been pleased. But they did not do what they did for the profits.

Indeed, for Aristotle and Confucius, the purpose of life is moral self-growth. One's life assumes meaning by virtue of one's improvement of one's character.[27] One improves one's character by individual acts of moral choice. Morality, or the moral person, is defined by the choices that one makes. In Confucius' *Analects*, it is written, 'The gentleman understands what is moral. The small man understands what is profitable.'[28] Ultimately, all of these life choices that one makes along life's way lead to one's moral character.

The entire purpose of man's life, for Aristotle, since the life of pure contemplation is beyond most men, is in choosing moral acts to perform. Society exists for this very purpose. For the famous Athenians in the Golden Age of Greece, they differentiated themselves from the Phoenicians with their dismissal of this nation of merchants as 'loving only money'.

Consider this famous passage from Sophocles', *Antigone:*

> *No thing in use by man, for power of ill,*
> *Can equal money. This lays cities low,*
> *This drives men forth from quiet dwelling-place,*
> *This warps and changes minds of worthiest stamp,*
> *To turn to deeds of baseness, teaching men*
> *All shifts of cunning ...*

This is a different universe than the universe that is formed by the logical consequences of the theory of Adam Smith. One obtains a moral society by the performance of moral actions. One cannot obtain a moral society by the pursuit of self-interest. It must be said that while Professor Smith was a Professor of Logic before he became a Professor of Moral Philosophy, there appears to be a better logic in the arguments of Aristotle and Confucius than in those of Adam Smith. For Smith the aggregate of self-interest leads to the good of all. For Aristotle and Confucius, the aggregate of moral actions leads to the good society.

5. Making Two Faces into One

We cannot theorize morality on the one side and economic behavior on the other and hope to patch them together in some fashion. The ancient Greeks and Chinese (and I only choose these two peoples as an example not to prove that they were unique) put forth a view of humankind that held that planned ethical motivation was the motivation for living. Not so with Adam Smith. For Adam Smith, economic motivation is self-aggrandizement. In terms of economic action, ethics comes into view only as an extrinsic and unplanned outcome. He does, of course, provide separately for an ethics of sympathy and a famous one at that. But the problem is that it is not ultimately consistent with his description of what is best for society as a whole.

There is another point to be considered. When self-interest is served, as Plato well knew, the appetites grow. This is one of the problems with proclaiming self-interest to be the centerpiece of human motivation. For what is to draw the line between self-interest and greed? If one hundred thousand dollars is good as my annual income, why not two hundred thousand? And if two hundred thousand is good, why not three hundred thousand? And so on. Once greed enters the picture, the distinction between self-interest and greed, if there is one at all, begins to break down. What is to place the limits on greed? As we have seen in the creation of multi-billionaires in both the West, in China and other countries, there are no limits on greed.

The problem is not how to restrict greed, for that is well nigh impossible. The problem is with having made greed the essential motivator for economic action in the first place. The maximization of profit is the economic manifestation of the psychological motivation of greed. It is not likely by starting out with selfishness as the essential motivator that one is going to reach unselfishness at the end of this road. Why would anyone attempt to posit a theory of human motivation based on greed? Granted that it is nominally self-interest and not greed, the distinction between self-interest and greed is a slippery slope. The ancient Greeks and the ancient Chinese knew better than this. If one begins with self-aggrandizement, how can one end with equality for all? Has there ever been such an egregious *non-sequitur* in the history of humankind to rival this monstrous stroke of illogic?

Joan Robinson, the Cambridge economist has argued that 'The emergence of industrial capitalism required the existence, on the one side, of a proletariat – that is many families who had no rights in land or possession of means of production, so that a great number of individuals were available to be employed for wages – and, on the other side, a few families with large accumulations of wealth which could be used to employ them in such a way to yield profits. I do not think that any academic economist could deny this obvious fact, but they have elaborated their theories in such a way to conceal it.'[29]

And again, 'The nature of accumulation under private enterprise necessarily generates inequality and is therefore condemned to meeting the trivial wants of a few before the urgent needs of the many.'[30]

But she could have found such a statement in the famous book by the father of economics, Adam Smith:

'Wherever there is great property, there is great inequality. For one very rich man, there must be at least five hundred poor, and the affluence of the few supposes the indigence of the many. … It is only under the shelter of the civil magistrate that the owner of that valuable property … can sleep a single night in security.'[31]

Adam Smith himself was not happy with the untrammeled pursuit of profits:

'Our merchants and manufacturers … say nothing concerning the bad effects of high profits. They are silent with regard to the pernicious effects of their own gains.'[32]

And, 'No society can surely be flourishing and happy, of which the far greater part of the members are poor and miserable.'[33]

But the problem with Smith is that it is his very own theory that is, I argue, the source of the rationalization of and therefore to some extent the toleration of the production of inequity in society. As unhappy as he might be with the results, he provides, along with his contemporary apologists, considerable self-justification for its continuance. In addition, because of his noteworthy incursion into ethics, he might also be said to be responsible for an additional burden, the division of the human being into two parts, the ethical man and the economic man. Adam Smith may be held accountable for the divided self.[34]

If a moral philosophy is designed as a guide to human action, we should not have a moral philosophy that is at odds with an economic philosophy. Otherwise, we create a divided self, we condemn one of these philosophies to the dustbin or we are hypocrites. If we base our economic theory on self-interest, then corporate social responsibility is something which comes afterwards. It is either something that flows naturally, as in Adam Smith, as a necessary result or intrinsic result of the functioning of the economic system, or it is an ethical add on, an extrinsic function corporations must perform as an obligation of their economic power, a kind of economic version of *noblesse oblige*. In the second sense, there is no logical connection between the corporation's pursuit of profit and its responsibility or accountability for the plight of society. Ethics is an obligation incurred by success. Ethics becomes a kind of penalty imposed upon the successful business venture.

Evidence indicates that rather than equitable distribution occurring as a result of the pursuit of private wealth, the opposite occurs. Thus, the first approach, the profits approach, results in inequity. This is not surprising since it would appear to be illogical if the pursuit of greed, that is, more than one needs, should result in economic justice.

But there is an additional factor as well, what may be called the unhealthy mixture. This view bears some similarity with the interesting analysis of Luk Bouckaert. While he does not use the term 'unhealthy mixture', he argues that genuine ethics may be crowded out by an ethics co-opted by management.[35] In the argument presented by the present author, when ethics is viewed as an extrinsic responsibility, then the original business ingredient is tainted by being viewed as unethical. This creates an unethical self-image of the business person. He or she becomes ethical only when performing extrinsic ethical acts. This, is, although better than a business that has no interest in social accountability at all, an unhealthy mixture. In addition, by identifying the ethical component in the extrinsic feature, what is lost is the possibility of a greater ethical contribution that could be made by considering ethics to be a part or even a whole of the initial business concern.

What else is lost, and this may be even more pernicious, is that in daily life, if we think that every man and woman is out for his or her own self-interest, then every man or woman is met with an initial and well-justified mistrust. The self-interest model of economics creates distrust rather than trust as an initial starting point for human interaction. This is a serious reduction in the quality of interpersonal relations. How do we know when our dearest friend will slip a knife into our back when economic advantages dictate the necessity of such an action?

6. Value Creation as a Theoretical and Practical Foundation for Economics

Our new theory of economics or value creation is the notion that all values, aesthetic, social and ethical are to be part and parcel of one's motivation for action in life and profit is to be the secondary but not logically consequential effect. We reverse the entire equation. Action is for value creation, not for profit. Profit is the side-effect, not the goal. We build ethics and aesthetics into the business in the first place.

We may venture for a definition of the new economics the following definition:

Value based economics may be defined as the ownership or the use of capital investment, labour or land to produce a product or provide a service that fills a value based social need or creates a positive social value without creating a disvalue which is of greater harm than the value produced. The more that the good or service contributes to the social value of the underdeveloped world, the greater the value of the good or service that is provided.[36]

In a value based economics, all car manufacturers would cease making cars that ran on oil product derivatives and all car manufacturers would sell only cars that ran on electricity or were hybrids. Since it would be easy to buy a car that was a hybrid, no surcharge for scarcity could be attached to a hybrid car. The buyer would be able to buy a hybrid easily. The seller's profit margin might be reduced (or it might increase given the volume of sales). But, the main result would be that buyers would spend less money on gas; the environment would become cleaner; the price of oil would become reduced because of decreased demand.

In the above scenario, buyers would be able to purchase and use vehicles for transport on the basis of saving money and saving the environment. A value choice would be available to them. *A non-self sacrificial value choice becomes available to buyers when the seller's motivation is the production of something that creates positive value rather than something that either reduces value or creates disvalue. Value creates value; disvalue creates disvalue.* When the seller creates disvalue, e.g., a car that pollutes and causes an unnecessary depletion in available assets in the buyer, the non self-sacrificing buyer might be impelled to make up those funds by creating more disvalue in turn. When the seller creates value, e.g., a car that does not pollute and causes an increase in available assets in the buyer, the buyer can create value with the increase in available assets.

In a value-based economics, the car maker will want to make as many cars available as possible at the lowest possible cost to the consumer while not making any sacrifices in quality. The manufacturer will create value for others. One result of this, assuming everyone buys only hybrid cars, is that the price of oil would drop dramatically. This would mean that the oil rich nations would lose some of their comparative advantage of wealth. The creation of value automatically creates a redistribution of wealth. *The creation of needed new value in one part results in the reduction of excess value in another part.* While it may appear that the reduction of value in the oil rich countries is a loss, it is not exactly a loss; it is a reduction of excess. The creation of needed value where it did not exist balances excesses in parts where value is not needed (or is superfluous).

On the other hand, in a profit-based economics, every business attempts to make its greatest profit by keeping wages, quality and availability of the good or service it produces to the lowest possible level compatible with sales. Quality and availability of the product or service must always be sacrificed for profit. *There is an inverse relationship between quality and/or availability and profit. Profit inherently works in an inverse relationship to quality and universal distribution.* This is the law of disequilibrium.

Country C sells goods at a lower price to country A because of lower wages in country C. This creates value and disvalue for country A. It creates value because country A can spend less for its goods that it needs. It creates disvalue because country A no longer manufactures such goods for its own use or for export to country B. This creates value for country C. Country C can sell a monopoly of goods to country A. Eventually, however, country C must raise its salaries and it can no longer supply goods to country A. Now, country D sells goods at a lower price to country A. Country C begins to suffer. What was a value to country C eventually becomes a disvalue. *Whatever brings value to oneself at a price of disvalue to another eventually brings disvalue back to itself.*

If the nature of man is such that man's motivation for buying and selling is the making of profit, human beings will attempt to restrict production and supply of needed items to the quantity consonant with the greatest profit margin. If someone is manufacturing cars which do not rely upon oil, for example, one would restrict production of this kind of car to an amount that would enable the owner or manufacturer to sell that number of cars that will bring the highest amount of profit.

A system which is based on justice will result in justice. A system which is based on greed will result in greed. Adam Smith devised a system in which the basic motivator of greed was supposed to result in equality or at least equity. It is not possible to go from inequality to equality. If one starts from greed and envy, one can only succeed in producing greed and envy. *Like produces Like*. Or, *Like can only come from Like*. This is the general principle that governs value creation. Adam Smithian economics is inherently flawed. One cannot produce distributive justice from self-aggrandizement. The invisible hand is worse than an invisible God. With an invisible God, Abraham could at least negotiate.

When one acts out of greed or fear and pursues profit at the expense of loss for someone and distributive justice for many, one always wants more.[37] One is never satisfied. This is why for Plato and Aristotle, the pursuit of wealth could not result in happiness. Since happiness was the goal or at least the natural state for human beings, neither Plato nor Aristotle could have selected the rational economic man as the model for human action. Economics must be based on a true and natural function of humanity.

7. The Unity of the Ethical Man and the Businessman

Our life should be one whole. We should not need to make a dichotomy between our business decisions and our ethical decisions. Our life should be of one piece. Given our ethical nature, we should not have to go against it. It is true that the way human institutions have been set up, particularly with profit based economics, it is difficult to combine ethics with business. At best one can minimize profit and attempt to make profit through industries that are socially contributive. But still, even if a business is socially contributive, the way it makes its profits may be creating losses for others. And if one minimizes one's profit, one places oneself at a disadvantage to others. And one's self is just as important as other selves. It is a sad lot to live and prosper in a profit based economic system and one can only accomplish this by burying one's head under the ground. A noble attempt to rectify this situation is Prakash Sethi's replacement of corporate social responsibility with his well argued notion of corporate accountability.

With Mencius our nature is to be compassionate to other beings. It is our core compassionate nature, not our desire to look good in the eyes of others, that is the origin of our morality. That being the case, why should this nature not be our guide in all of our activities? Surely our compassion does not stop at the door of our business. If it does, we have placed an artificial barrier between our nature and our business action. If we cannot satisfy our ethical nature in our business transactions, it follows that our business behavior is *unnatural.* Judged in this way, Adam Smith's counsel for each individual to pursue her or his self-interest (albeit while not doing anything unethical), is *an unnatural guidance system.*

8. Capitalism as Inherently Unethical

The whole point is that capitalism constrains us to be unethical. Not only that, capitalism constrains us to be unnatural. Indeed, since it extols self-interest at the expense (read, competition with) the other, it daily trains us to become unethical. This being the case, how difficult it must be to summon up ethical behavior in non-business situations so habituated as we must be to immoral behavior!

Indeed though much has been written about how ethical trust is the basis for economic behavior, if one believes that the entire point of economic behavior is to take advantage of, that is profit over the other, then one's attitude towards other human beings cannot be one other than arrogance, distrust, fear and contempt or indifference. Profit based economics breeds arrogance in the profiteer, ethical distrust and fear of the profiteer and contempt or indifference towards the victim. Profit based economics cannot help but breed ethical distrust, fear and contempt (or at least indifference) as social properties.[38] This is the ethical legacy of profit based economics.

By following a self-interest and profit based economics, one inevitably shapes one's character, but in this case, one shapes an unethical character. This unethical character, molded by the daily pursuit of profit, must come to battle with one's ethical impulses, nourished on those rare occasions when one is not seeking profit but is engaged in "pure" ethics. In addition, since the success of society is based on the existence of classes, one daily breeds an unethical society.

How can one cast away one's immoral net on these special, unnamed moral occasions, and just as quickly and readily wrap it around our shoulders when entering the business arena? We become, like the Mafia, able to discriminate between actions for family and the murderous actions that are not personal, but only business.

Unlike the Mafia, whose hit men must only upon occasion practice the unethical act of murder, in the world of business, it is every decision that is calculated to ensure that it is to one's advantage. When one is day by day, hour by hour, minute by minute, performing cost-benefit analyses to determine which course of action to take, is it really possible to throw off this calculating brain and embrace humankind in a compassionate hug? Or, are even these rare and discrete acts calculated as well so that they do not overly interfere with the daily progress and pressure of our business and professional life. These ethical actions are perforce relegated to the back stream of life, to be practiced perhaps in retirement or on Christmas Day, but for the rest of the year or one's career, surely to make up only a tiny minority of our actions, if any.

Now it is true that Adam Smith does frequently say that one must be ethical when carrying out one's business; one must not practice any deceit, for example. One is not entitled to use any means, fair or foul, to increase one's profit margin. One must, for Adam Smith, observe basic ethical amenities.

9. Conclusion

The whole point is that a system founded upon self-interest is inherently unethical as a system, not by its abuse. *It is unethical in its proper use. We teach our children to share with each other because it is unethical to take for oneself and the neglect the other.* The entire doctrine, much lauded, of comparative advantages, is an explicit articulation of this unethical ideal. *My comparative advantage must be your comparative disadvantage.* If it were not, it would not be my comparative advantage. It is a simple as that.

There are those who would even argue that this is the price of economic efficiency. In these terms, poverty would always be present; it would even be required. It might even be glorified. It reminds one of Johnson's remark that 'Sir, the great deal of arguing which we hear to represent poverty as no evil shows it to be evidently a great one. You never knew people labouring to convince you that you might live very happily upon a plentiful fortune.'[39]

To conclude, it appears as if Adam Smith is in conflict with himself. His view of humans as ethical beings collides with his view of human beings as pursuing self-interest as presented in his *Wealth of Nations*. You cannot have it both ways. But, most of us would like it that way. That is in fact the entire motivation for the discipline of business ethics. If business is self-interested it must, to satisfy our ethical urges, be disciplined by ethics. This is where we are today.

Economics cannot exist for the sake of itself: it must exist for the sake of something else. Our work, our labours, our investments, all exist for the sake of something higher, of a higher end. As Joan Robinson once asked, 'Here we come upon the greatest of all economic questions, but one which is never asked, what is growth for?'[40] Herein, the answer is given. Proper growth serves the end of the creation of positive value and the reduction of excess value, disvalue or negative value.

Growth is a quantitative concept. It cannot comprehend quality. One must consider the quality of the growth. The growth in the production of automobiles increases traffic and both noise and air pollution. Contrariwise, shrinkage cannot be automatically construed to be a disvalue. The shrinkage in the automobile industry could have qualitatively favorable results. It might result in the decrease in noise and air pollution. It might result in the growth of public transportation. The growth in public transportation might result in greater convenience and less cost.

Growth is not an end in itself. Growth, in and of itself, cannot be said to be intrinsically valuable. It depends upon the content of the growth and the effects of the growth. With respect to content, if the growth is the growth of a cancer, growth in this instance is a disvalue rather than a value. With respect to effects, if the growth, for example the gross national product of one country, increases at the cost of the growth national product of another country, or multiple countries, then the growth of the gross national product of the first country cannot be said to be an unqualified value. It may, in fact, create more disvalue than value, all countries and all peoples to be considered.

Joan Robinson was correct is raising the question of what is growth for. If growth were the end-goal of economic activity, economics would end with a question, not an answer. It would end with the question Joan Robinson asked, what is growth for?

By making value, as contrasted with disvalue, the end-goal, growth is given a purpose and a direction. Economics is able to become a moral economics, an end which Adam Smith desired, but could not reach without a basic conflict within his system of thinking. By making value creation the purpose of economics, economics has an end-goal which also serves the purpose of self-correction. Economics has a built-in moral compass as well. As soon as it is sensed that economic activity possesses the potential to create disvalue, one realizes that economic activity is misdirected. One would not, for example, offer a 100% loan to a house buyer without considering the potential disvalue such a loan could create. The built-in consideration of ethical consequences has the potential of saving countries from huge financial disasters as well.

Economic activity is given a higher purpose. Its purpose is to serve the higher ends of reducing negative value, that is, moral harm, and increasing positive value, that is, producing moral good. It is entirely likely that a general material or financial benefit, one that is distributed more equitably throughout the world, will also be the result. This follows from the general principle that governs value creation, like produces like, or, like can only come from like. An answer to the astute question that the Cambridge economist Joan Robinson stated to be the 'greatest of all economic questions, but one which is never asked, what is growth for?' is hereby asked and answered by an ethically based economics. Economics should not end with the question Joan Robinson asked. It should begin with it. The proper beginning of economics is to ask the question, what is growth for? The proper beginning of economics is the construction of the theory of the economics of value creation.

Notes

1 As to whether ethicists or philosophers are competent to comment on economics, the reader must indulge the present author in a lengthy quotation from John Rae's, *Life of Adam Smith* in which he recounts the question being raised by a Professor of Moral Philosophy, a Sir John Pringle who 'remarked to Boswell that Smith, having never been in trade, could not be expected to

write well on that subject any more than a lawyer upon physic, and Boswell repeated the remark to Johnson, who at once, however, sent it to the winds. "He is mistaken, sir," said the Doctor; "a man who has never been engaged in trade himself may undoubtedly write well upon trade, and there is nothing that requires more to be illustrated by philosophy than does trade. As to mere wealth – that is to say, money – it is clear that one nation or one individual cannot increase its store but by making another poorer; but trade procures what is more valuable, the reciprocation of the peculiar advantages of different countries. A merchant seldom thinks but of his own particular trade. To write a good book upon it a man must have extensive views; it is not necessary to have practiced to write well upon a subject."' *Cf.*, Rae, J. (1895, 1965). *Life of Adam Smith* (p. 288). New York: August M. Kelley. It should be added to this that not only was Smith not in trade but his study in Glasgow was of moral philosophy and his later study at Balliol was of history, philosophy and Latin and Greek literature. *Cf.*, Lothian, J. M. (Ed.) (1963). *Adam Smith's Lectures on Rhetoric and Belles Lettres* (p. xiv). London: Thomas Nelson and Sons, Ltd.

2 For a comprehensive, fair and well balanced view of the different interpretations of Adam Smith and her own well-argued view, *cf.* the arguments of Werhane, P. (1991). *Adam Smith and His Legacy for Modern Capitalism.* New York: Oxford University Press.

3 Viner, J. (1965). Adam Smith. *International Encyclopedia of the Social Sciences* (p. 326). New York: Macmillan.

4 Schumpeter, J. A. (1954, 1960). *History of Economic Analysis* (Ed. E. B. Schumpeter) (p. 184). New York: Oxford University Press.

5 *Cf.*, No. 10 below.

6 *Cf.*, *History of Political Economy*, 3, 1971, p. 265. In his article on Adam Smith for the *International Encyclopedia of the Social Sciences*, Viner appears to present a softer view of Smith. The position I take is that no matter how Smith qualifies his views, e.g., that the desire to better our conditions should be disciplined by internal or governmentally imposed justice, the point is that the logic of Smith's argument is that such discipline of the desire to materially improve ourselves must be counterproductive to the goal of the improvement of the wealth of the nation. *Cf.*, Viner, J. (1965). Adam Smith (pp. 322–329). *International Encyclopedia of the Social Sciences.* New York: Macmillan.

7 *Cf.*, Irwin, D. A. (Ed.) (1991). Jacob Viner (p. 3). *Essays on the Intellectual History of Economics.* Princeton: Princeton University Press.

8 *Cf.*, Evensky, J. (2005). *Adam Smith's Moral Philosophy.* Cambridge: Cambridge University Press.

9 *Cf.*, Morrow, G. R. (1923, 1973). *The Ethical and Economic Theories of Adam Smith.* Clifton: Augustus M. Kelley.

10 *Cf.*, Viner, J. (1972). *The Role of Providence in the Social Order* (p. 80). Philadelphia: American Philosophical Society.

11 Stigler, G. (1976). The Successes and Failures of Professor Smith. *Journal of Political Economy*, 84 (6), pp. 1199–1213. Interestingly, Viner considers the famous idea of 'equilibrium' to be a normative idea. It is not certain how he means this. He might mean that it is not an analytic idea. He states, "If 'analytical' as a eulogistic term is to be interpreted strictly in terms of degree of rigor, internal consistency, and close analogy to abstract mathematical operations ...Schumpeter's verdict [that there were no new single *analytic* ideas of principles in the *Wealth of Nations*]... is difficult to challenge." *Cf.*, Viner, J. (1965). Adam Smith (pp. 327). *International Encyclopedia of the Social Sciences.* New York: Macmillan.

12 Grampp writes, 'The famous invisible hand of *The Wealth of Nations* is nothing more than the automatic equilibration of a competitive market,' Grampp, W. D. (1948). Adam Smith and Economic Man. *Journal of Political Economy*, (4), p. 334. Cited in Evensky, J. (2005). *Adam Smith's Moral Philosophy* (p. 246). Cambridge: Cambridge University Press. One may also trace this to the fundamental idea of the *Tableau economique* of Quesnay whose metaphor of the circulation of wealth in human societies was modeled after the circulation of blood in the human body, a metaphor which was to dominate the notions of the physiocrats. Smith even meant to dedicate his *Wealth of Nations* to Quesnay had Quesnay been alive at the time. *Cf.*, Rae, J. (1895, 1965). *Life of Adam Smith* (p. 216). New York: Augustus M. Kelley. Here, in passing one must pay tribute to the power of the salons in France and thus to the association of ladies with ideas, which might account for the decline of intellectuality in the twenty-first century in which salons are no longer in vogue, for conversations with Turgot, Quesnay's famous disciple, took place in the salon of Mademoiselle de l'Espinasse. *Cf.*, *Ibid.*, pp. 201–204. Rae also recalls that Smith was a regular guest of the Duchesse d'Enville, herself a grand-daughter of the celebrated la Rochefoucauld, and who was popularly supposed to the inspiration of Turgot's ideas. In addition, Smith sat at table with Mirabeau and the young Duc de la Rochefoucauld, who was a disciple of Quesnay and friend of Turgot. *Cf.*, *Ibid.*, p. 192. Mirabeau famously declared the *Tableau* one of the three greatest inventions of the human spirit along with printing and money.

13 This seems to be the direction of Jacob Viner's argument in his *The Role of Providence in the Social Order*, Philadelphia: American Philosophical Society, 1972. An interesting primary evidentiary source of this idea is an opinion of Smith's cited by the early biographer of Smith, Dugald Stewart , was that in 1752 or 1753 Smith delivered lectures at Glasgow that contained the funda-

mental principles of the *Wealth of Nations* and in virtue of a paper of Smith's that Stewart had in his possession (that escaped Smith's own fire but later was to succumb to Stewart's own son's fire), Stewart offered the following quotation of which I reproduce but a portion: 'Projectors disturb nature in the course of her operations on human affairs, and it requires no more than to leave her alone and give her fair play in the pursuit of her ends that she may establish her own designs. Little else is required to carry a state to the highest degree of affluence from the lowest barbarism but peace, easy taxes, and a tolerable administration of justice; all the rest being brought about by the natural course of things.' Cited in Rae, J. (1895, 1965). *Life of Adam Smith* (pp. 62–3). New York: Augustus M. Kelley. While at first glance this quotation would not seem to support the idea of a providential order, upon reflection one may consider that nature is purported to have a natural direction towards economic affluence. Whether providential or natural is not to the point. {Morrow identifies the concept of Nature with the Divinity in Smith, citing *Moral Sentiments*, Pt. II. Sec. II, chap. iii. *Cf.*, Morrow, G. R. (1966). Adam Smith: Moralist and Philosopher. In J. M. Clark et al. (Eds), *Adam Smith, 1776–1926* (p. 171). New York: Augustus M. Kelley} What is to the point is that prosperity is perceived of as the result of not disturbing what will naturally occur. This differs not in substance from the doctrine of the invisible hand though in this instance the invisible hand (or no hand at all) has more to do with obtaining prosperity without regard to distribution. While self-interest is also not mentioned here, it still seems noteworthy that Smith thought that the highest degree of affluence was essentially linked to a notion of non-interference.

14 Smith, A. *The Wealth of Nations*, Book I, Chapter VII, p. 63.

15 Sen, A. (1996). Does Business Ethics Make Economic Sense? In T. Donaldson & P. H. Werhane (Eds), *Ethical Issues in Business: A Philosophical Approach* (fifth edition, p. 17). Upper Saddle River, New Jersey: Prentice Hall.

16 For an extended argument, *cf.*, Allinson, R. E. (2004). Circles within a Circle: The Condition for the Possibility of Ethical Business Institutions within a Market System. *Journal of Business Ethics*, 53: 17–28.

17 Sen, A. (1997). *On Economic Inequality* (p.110). Expanded edition with a substantial annex by James E. Foster and Amartya Sen. Oxford: Clarendon Paperbacks.

18 Strange though it may seem, Adam Smith did not even found his ideas of the motivation for trade on self-interest. According to Jacob Viner, 'Adam Smith has puzzled many commentators by his attribution of the origin of commerce to a subrational propensity to truck and barter, rather than to a rational pursuit of economic benefit.' *Cf.*, Viner, J. (1972). *The Role of Providence in the Social Order* (p. 47). Philadelphia: American Philosophical Society.

19 Smith, A. (2000). *The Theory of Moral Sentiments*, Part IV, Chapter I, pp. 264–5. Amherst, New York: Prometheus Books.

20 Smith, A. *The Wealth of Nations*, Book IV, Chapter II, pp. 569–570. (Bantam).

21 Smith, A. *The Wealth of Nations*, Book I, Chapter X, p. 138 (Bantam).

22 Jacob Viner, 'Adam Smith and Laissez Faire,' John Clark, et al. (Eds), *Adam Smith, 1776–1926*, Augustus M. Kelley, 1966, p. 130. Cited in the charmingly written, Marjorie Ann Clay, 'Private Vices, Public Benefits': Adam Smith and the Concept of Self-Interest', William R. Morrow and Robert E. Stebbins, (Eds), *Adam Smith and the Wealth of Nations*, 1776–1976, Proceedings of the Bicentennial Conference, Eastern Kentucky University, 1976, p. 51. Patricia Werhane refers to Marjorie Ann Clay's chapter as 'providing a good summary of these [the Adam Smith problem] problems.' Patricia Werhane, *Adam Smith and His Legacy for Modern Capitalism*. Oxford: Oxford University Press, 1991, p. 183, No. 37.

23 Smith, A. *The Theory of Moral Sentiments*, Part I., Section I., Chapter I., p. 3.

24 Even Viner at times takes up the cudgel of Smithian apologists. In his Adam Smith article for the *International Encyclopedia of the Social Sciences,* Viner states that Smith 'would probably have demonstrated that the apparent inconsistencies were often not real ones, but were merely the consequences of deliberate shifts from one partial model to another.' *Cf.,* Viner, J. (1965). Adam Smith. *International Encyclopedia of the Social Sciences* (p. 323). New York: Macmillan. The problem remains that if one's interest is for the prosperity of nations, there is no contest as to which model is to be chosen. That necessarily means the ignoring of the other model which is the ethical model.

25 Confucius, *Analects*, Book IV, 10, 16; Book XIV, 12; XVI, 10; Book XVII, 23.

26 For a preliminary account of what is offered here, *cf.*, Allinson, R. E. (2004). The Birth of Spiritual Economics. In L. Zsolnai (Ed.), *Spirituality and Ethics in Management.* Dordrecht, Boston, London: Kluwer Academic Publishers. Issues in Business Ethics, Vol. 19, 2004, pp. 61–74.

27 Confucius, *Analects*, Book II, 4.

28 Confucius, *Analects*, Book IV, 16.

29 Robinson, J. (1980). *Collected Economic Papers*, Vol. V, p. 291. Cambridge: MIT Press.

30 Robinson, J. (1980). *Collected Economic Papers*, Vol. V, p. 30. Cambridge: MIT Press.

31 Smith, A. *The Wealth of Nations*, Book V, Part II, p. 902.

32 Smith, A. *The Wealth of Nations*, Book I, Chapter IX, p. 137.

33 Smith, A. *The Wealth of Nations*, Book I, Chapter VIII, pp. 110–111.

34 One could well argue that the divided self is discoverable even within the limits of *The Theory of Moral Sentiments* for even there the good of society is accomplished precisely through the mechanism of self-interest. Logically speaking, some of the mechanism of moral sentiments, particularly that of altruism, would work against the good of society. If the rich did not pursue their self-interest, the poor would remain poorer.

35 Zsolnai, L. (Ed.) *Business Ethics Yearbook* 2006, The Debate Section.

36 Robert Elliott Allinson, 'Circles within a Circle: The Condition for the Possibility of Ethical Business Institutions Within a Market System,' *Journal of Business Ethics*,' 53: p. 26, 2004.

37 For an extended discussion of greed and other derivative emotions, *cf.*, Allinson, R. E. (2002). *Space, Time and the Ethical Foundations* (pp. 147–163). Aldershot, Burlington: Ashgate Publishers.

38 Luigino Bruni and Robert Sogden claim that Smith thinks that trust is a product of commercial society. For an interesting discussion, *cf.*, Bruni, L & Sugden, R. (2000). Moral Canals, Trust and Social Capital in the Work of Hume, Smith and Genovesi. *Economics and Philosophy*, 16, pp. 21–45.

39 This quotation from Boswell's *Journal* is to be found in J. Viner (1972), *The Role of Providence in the Social Order* (p. 105). Philadelphia: American Philosophical Society.

40 Robinson, J. (1980). *Collected Economic Papers*, Vol. V., p. 29. Cambridge: MIT Press.

Laszlo Zsolnai
Business Ethics Center
Corvinus University of Budapest

Buddhist Economics for Business

The paper explores Buddhist economics for transforming business toward a more ecological and human form. Buddhist economics is centered on want negation and purification of the human character. It challenges the basic principles of Western economics, (i) profit-maximization, (ii) cultivating desires, (iii) introducing markets, (iv) instrumental use of the world, and (v) self-interest based ethics. Buddhist economics proposes alternative principles such as (I) minimize suffering, (II) simplifying desires, (III) non-violence, (IV) genuine care, and (V) generosity. Buddhist economics is not a system but a strategy, which can be applied in any economic setting. Buddhist economics provides a rational, ethical, and ecological value background, which promotes happiness, peace and permanence.

1. The Conception of "No-Self"

Thomas Shelling characterized Western economics as an "ego-nomical framework". Western economics is centered on self-interest understood as satisfaction of the wishes of one's body-mind ego. Buddhism challenges this view by a radically different conception of the self, that is, "anatta", the "no-self". (Elster 1985)

Anatta specifies the absence of a supposedly permanent and unchanging self. What is normally thought of as the "self" is an agglomeration of constantly changing physical and mental constituents which give rise to unhappiness if clung to as though this temporary assemblage. The "anatta" doctrine attempts to encourage the Buddhist practitioners to detach themselves from the misplaced clinging to what is mistakenly regarded as self, and from such detachment (aided by moral living and meditation) the way to *Nirvana* is able successfully to be traversed.

Modern neuroscience supports the Buddhist view of the self. What neuroscientists discovered is can be called the selfless (or virtual self), "a coherent global pattern, which seems to be centrally located, but is nowhere to be found, and yet is essential as a level of interaction for the behavior". The non-localizable, non-substantial self acts as if it were present, like a virtual interface. (Varela 1999. pp. 53, 61.)

The Buddhist conception of selflessness has enormous implication for economics. In the works of Schumacher, E. F., Venerable, P., Payutto, A. and others Buddhist economics is emerging as a major alternative to the Western economic mindset. (Schumacher 1971; Payutto 1994; Zsolnai and Ims (Eds) 2006)

2. Minimize Suffering

When Western economics promotes doing business based on individual, self-interested, profit-maximizing way, Buddhism suggests an alternative strategy. The underlying principle of Buddhist economics is to *minimize suffering* of all sentient beings including human and non-human beings.

In more technical terms the suffering minimizing principle can be formulated that the goal of economic activities is not to produce gains but the decrease losses. This is an adequate strategy in the light of experimental decision research.

The *prospect theory* developed by *Daniel Kahneman* and *Amos Tversky* discovered the basic empirical features of the value function of decision makers. The central finding of prospect theory is that decision makers are more sensitive to losses than to gains. (Kahneman, D. and Tversky, A. 1979) Experiments show that the ratio of the slopes in the domains of losses and gains, the "loss aversion coefficient", might be estimated as about 2:1. (Tversky and Kahneman 1991, 1992)

Because humans (and other sentient beings) display loss-sensitivity it does make sense trying to reduce losses rather than trying to increase gains. Losses should not be interpreted only in monetary terms. Also they should not apply only to humans. Suffering, that is the capability of experiencing losses, is *universal* in the realm of natural and human kingdom.

3. Simplifying Desires

Western economics cultivates desires. People are encouraged to develop new desires for things to acquire and for activities to do. The profit motive of companies requires creating more demand. But psychological research shows that materialistic value orientation undermines well-being. "People who are highly focused on materialistic values have lower personal well-being and psychological health than those who believe that materialistic pursuits are relatively unimportant. These relationships have been documented in samples of people ranging from the wealthy to the poor, from teenagers to the elderly, and from Australians to South Koreans." These studies document that "strong materialistic values are associated with a pervasive undermining of people's well-being, from low life satisfaction and happiness, to depression and anxiety, to physical problems such as headaches, and to personality disorders, narcissism, and antisocial behavior." (Kasser 2002, p. 22.)

Psychologists call "auto-projection" the mechanism through which people seek to satisfy their desires. It is a *looser strategy* whether or not people achieve their desired goals. When they are not able to reach the goals they envision, they attribute their continuing dissatisfaction to their failure to reach the alleged corrective measures. When they succeed in attaining their goals, this usually does not bring what they hoped for and their feeling of discomfort are not relieved. So striving for satisfying desires never bring people the fulfillment they expect from it. (Grof 1998, p. 207.)

The Buddhist strategy suggests not to multiply but to simplify our desires. Above the minimum material comfort, which includes enough food, clothing, shelter, and medicine, it is wise to try to reduce one's desires. Wanting less could bring substantial benefits for the person, for the community, and for nature.

Buddhism recommends moderate consumption and is directly aiming at changing one's preferences through meditation, reflection, analyses, autosuggestion and the like. French economist *Serge-Christopher Kolm* developed a formal model to treat consumption and mediation together. (Kolm 1985)

In a simplified form his model is as follows. Let "u" represents one's *well-being* (or "sukkha"). Let "c" and "tm" represent consumption and meditation. These variables are linked by the relation u = u (c, tm).

The acquisition of consumption goods takes time, because labor is involved in producing them or needs to earn money to buy them. Let this length of time be "ta". The quantity of c is an increasing dependent variable of this, so c = c(ta).

We then have u = u [c(ta), tm]. Time should be divided between working for consumption and meditation. What is the optimal allocation between these two activities? The Buddha says that the optimum is some meditation to lower the desire for consumption and to be satisfied with less, and some consumption and thus to work that it entails. This is the "Middle Way". In economic terms this means "the marginal productivity of labor involved in producing consumption is equal to the marginal efficacy of the meditation involved in economizing on consumption without altering satisfaction". (Kolm 1985, pp. 240–242)

Desiring less is even fruitful in the case of money. Western economics presupposes that more money is better than less money. But, getting more money may have negative effect. Overpaid employees and managers do not always produce high-level performance.

Being under financed might be beneficial for a project. If people have smaller budget they may use the money more creatively and effectively. *Jesus* had no budget at all for financing his mission.

4. Practicing Non-violence

Western economics aims to introduce markets whatever social problems should be solved. *Karl Polanyi* called the whole process of marketization as "The Great Transformation" by which spheres of society became subordinated to the market mechanism. (Polanyi 1946) In the age of globalization we can experience this marketization process in a much larger scale and in a more speedy way than ever.

Market is a powerful institution. It can provide goods and services in a flexible and productive way, however it has its own limitations. Limitations of the market come from *non-represented* stakeholders, *under-represented* stakeholders, and *myopic* stakeholders.

Primordial stakeholders such as nature and future generations are simply not represented in the market because they do not have a "vote" in the terms of purchasing power. They cannot represent their interest in supply and demand. Other stakeholders such as the poor and marginalized people are under-represented because they do not have enough purchasing power to signal their preferences in the market. Finally, stakeholders who are well represented in the market because they have enough purchasing power, often behave in a myopic way, that is, heavily discount values in space and time. Market prices usually show the values of the strongest stakeholders and favor preferences here and now. Because of these inherent limitations the market cannot give a complete, unbiased direction for guiding economic activities. (Zsolnai and Gasparski (Eds) 2001)

Non-violence (called "ahimsa") is the main guiding principle of Buddhism for solving social problems. It is required than an act should not cause harm to the doer and the receivers. Non-violence prevents doing actions directly causing suffering for oneself or others and urges to find solutions by a participative way.

The community economy models are good examples. Communities of producers and consumers are formed to meet the needs of both of them at the lowest cost and reduced risk by a long-term arrangement. (Douthwaite 1996)

Community supported agriculture is the prime example of community based economic activities. Its essence is simple: a group of people agrees to buy in advance, shares of a farmer's harvest of food grown in an ecologically sound manner. It is a small-scale system whose central decision making body is the group of the farmer and the consumers. Community supported agriculture adopts a long-term perspective, de-commodify food and land, and reject monoculture and chemicals. Community supported agriculture strives to foster trust, to build value-community and to bring people closer to the land and the farm. (Dyck 1994)

Achieving ecological sustainability and non-violence requires altering the underlying structure of dominating configurations of modern business. This means de-emphasizing profit maximization and market systems and introducing small-scale, locally adaptable, culturally diverse way of substantive economic activities.

5. Genuine Care

According to the famous saying of *Oscar Wilde*: economists know the price of everything and the value of nothing. In Western economics the value of an entity (be it human being, other sentient being, object or anything else) is determined by its marginal contribution to the production output. The problem with this instrumental approach is that it generates the worst response from the parties involved. To get the best from the partners requires taking genuine care in their existence.

Robert Frank developed five distinct type of cases when socially responsible organizations are rewarded for the higher cost of caring. (Frank 2004)

(i) Opportunistic behavior can be avoided between owners and managers.

(ii) Getting moral satisfaction employees are ready to work more for less salary.

(iii) High quality new employees can be recruited.

(iv) Customers' loyalty can be gained.

(v) The trust of sub-contractors can be established.

Caring organizations are rewarded for the higher costs of their social responsible behavior by their ability to form commitments among owners, managers and employees and to establish trust relationships with customers and sub-contractors.

6. Generosity

There is a place for ethics in Western economics, however a little one. The Western economic man is allowed to consider the interest of others only if it serves his or her own interest. The *self-interest based*, opportunistic approach to ethics often fails.

Luk Bouckaert states taht the EU document of CSR policies is written under the veil of a rational and technocratic conception of ethics. Also, he explains that this conception of CSR is unable to overcome *opportunism* in business and politics. He argues that by reducing ethics to a functional and instrumental management concept we lose something vital. We are crowding out genuine moral feelings and genuine moral commitment, substituting them for rational and technocratic management tools. This substitution fails (Bouckaert 2006).

The *ethics management paradox* is the following. By creating new regulations to temper opportunistic behavior in and among organizations, we might temper the symptoms but often reinforce the underlying roots of opportunism. We introduce economic incentives like benefits, such as premiums or tax relief for those who respect the new regulations, but by doing this, we substitute moral feelings for economic calculations. Preaching moral concepts such as trust, responsibility or democracy on the basis of calculative self-interest or as conditions of systemic functionality opens the door for suspicion and distrust because calculations and systemic conditions can easily be manipulated. When the fox preaches, guard your geese. Bouckaert warns that the more economic democracy can be sustained by a rational and economic discourse, the more it risks crowding out the spiritual and moral commitment, which is a necessary condition for sustaining genuine entrepreneurship and stakeholding. Therefore we must put forward not only the question of how to make business ethics operational, but also the question of how to make it genuinely ethical. (Bouckaert 2006)

Generosity might work in business and social life because people are "Homo reciprocans." They tend to reciprocate what they get and often they give back more in value to the doer than he or she gave to them.

Ernst Fehr and *Simon Gaechter* designed a gift exchange game in which employer makes a wage offer with a stipulated desired level of effort from the worker. The worker may then choose an effort level, with costs to his or her rising in effort. The employer may fine the worker if his or her effort level is thought to be inadequate. The surplus from the interaction is the employer's profits and the worker's wage minus the cost of effort (and the fine, where applicable).

Self-regarding worker would choose the minimum feasible level of effort, and, anticipating this, the self-regarding employer would offer the minimum wage. But experimental subjects did not conform to this expectation. Employers made generous offers and workers' effort levels were strongly conditioned on these offers. High wages were reciprocated by high levels of efforts. (Bowles 2004. pp. 495–496)

7. Not a System but a Strategy

Buddhist economics represents a *minimizing framework* where suffering, desires, violence, instrumental use, and self-interest have to be minimized. This is why "small is beautiful" and "less is more" nicely express the essence of the Buddhist approach to economic questions.

Western economics represents a *maximizing framework.* It wants to maximize profit, desires, market, instrumental use, and self-interest and tends to build a world where "bigger is better" and "more is more". (Table 1)

Table 1
Characteristic of Western Economics and Buddhist Economics

Western Economics	Buddhist Economics
maximize profit	minimize suffering
maximize desires	minimize desires
maximize market	minimize violence
maximize instrumental use	minimize instrumental use
maximize self-interest	minimize self-interest
"bigger is better"	"small is beautiful"
"more is more"	"less is more"

Buddhist economics does not aim to build an economic system of its own. Rather it represents a strategy, which can be applied in any economic setting anytime. It helps to create livelihood solutions which reduce suffering of all sentient beings by practicing want negation, non-violence, caring and generosity.

Today's business model is based on and cultivates narrow self-centeredness. Buddhist economics point out that emphasizing individuality and promoting the greatest fulfillment of the desires of the individual conjointly lead to destruction.

Happiness research convincingly shows that not material wealth but the richness of *personal relationships* determines happiness. Not things but people make people happy. (Lane 1998) Western economics tries to provide people with happiness by supplying enormous quantity of things. But what people needs are caring relationships and generous love. Buddhist economics make these values possible by direct provision.

Peace can be achieved by non-violent ways. Wanting less can substantially contribute to this task and make it happen easier. *Permanence*, that is, ecological sustainability requires a drastic cut back of the present level of consumption and production globally. This reduction should not be an inconvenient exercise of self-sacrifice. In the noble ethos of reducing suffering it can be a positive development path for business. (Zsolnai and Ims (Eds) 2006)

References

Bouckaert, L. (2006). The Ethics Management Paradox. *Interdisciplinary Yearbook of Business Ethics,* Vol. 1, pp. 191–194.

Douthwaite, R. (1996). *Short Circuit. Strengthening Local Economics for Security in an Unstable World.* The Lilliput Press.

Dyck, B. (1994). Build in Sustainable Development and They Will Come: A Vegetable Field of Dreams. *Journal of Organizational Change Management,* Vol. 7, No. 4, pp. 47–63.

Elster, J. (1985). Introduction. In J. Elster (Ed.), *The Multiple Self* (pp. 1–34). Cambridge: Cambridge University Press.

Fehr, E. & Gaechter, S. (2000). *Do Incentive Contracts Crowd Out Voluntary Cooperation?* CEPR Discussion Paper, No. 3017. London: Centre for Economic Policy Research.

Frank, R. (2004). Can Socially Responsible Firms Survive in Competitive Environments? In Frank, R., *What Price the Moral High Ground? Ethical Dilemmas in Competitive Environments.* Princeton, Oxford: Princeton University Press.

Grof, S. (1998). *The Cosmic Game. Explorations of the Frontiers of Human Consciousness.* Albany: State University of New York Press.

Kahneman, D. & Tversky, A. (1979). Prospect Theory: An Analysis of Decision Under Risk. *Econometrica,* March, pp. 263–291.

Kasser, T. (2002). *The High Price of Materialism.* Cambridge, Massachusetts, London, England: MIT Press.

Kolm, S-C. (1985). The Buddhist Theory of 'No-self'. In J. Elster (Ed.), *The Multiple Self* (pp. 233–265). Cambridge: Cambridge University Press.

Lane, R. E. (1998). The Joyless Market Economy. In A. Ben-Ner, & Putterman, L. (Eds), *Economics, Values, and Organizations* (pp. 461–488). Cambridge: Cambridge University Press.

Payutto, P. A. (1994). *Buddhist Economics. A Middle Way for the Market Place.* http://www.geocities.com/Athens/Academy/9280/payutto.htm.

Polanyi, K. (1946). *The Great Transformation: The Political and Economic Origins of our Time.* London: Victor Gollancz Ltd.

Polanyi, K. (1971). *The Livelihood of Man.* New York: Academic Press.

Schumacher, E. F. (1973). *Small is Beautiful. Economics as if People Mattered.* Abacus.

Tversky, A. & Kahneman, D. (1991). Loss Aversion in Riskless Choice: A Reference-Dependent Model. *Quarterly Journal of Economics*. pp. 1039–1061.

Tversky, A. & Kahneman, D. (1992). Advances in Prospect Theory: Cumulative Representation of Uncertainty. *Journal of Risk and Uncertainty*. pp. 297–323.

Zsolnai, L. & Gasparski, W. (Eds) (2002). *Ethics and the Future of Capitalism*. New Brunswick, London: Transaction Publishers.

Zsolnai, L. & Ims, K. J. (Eds) (2006). *Business within Limits: Deep Ecology and Buddhist Economics*. Oxford: Peter Lang Publisher.

PART 2

Innovative Practices and Policy Reforms

Laura Albareda and
Josep Maria Lozano
ESADE Business School

Antonio Tencati and
Francesco Perrini
Bocconi University, Milan

Atle Midttun
Norwegian School of Management

The Role of Government in Corporate Social Responsibility*

Although the business sector lies at the core of the debate on Corporate Social Responsibility (CSR), other stakeholders participate in catalysing the various demands made by society on companies. The debate on the

* *This report has been prepared with the support of the European Academy of Business in Society (EABIS), as part of its Research, Education, and Training Partnership Programme on Corporate Social Responsibility. This Programme has been made possible by the financial support of EABIS' founding corporate partners, IBM, Johnson & Johnson, Microsoft, Shell, and Unilever.*

role of governments as promoters or drivers of CSR has been present since the early 1990s (Moon and Sochaki 1996; Aaronson and Reeves 2002a, 2002b; Fox et al. 2002; Zappal 2003) along with the first governmental policy to emerge on initiatives in the field. Nowadays, an important number of governments have taken policy initiatives promoting CSR.

In this multi-stakeholder debate, the public sector plays an important role in mediating between social agents. However, public sector initiatives throws up contradictions between stakeholders differing demands and expectations regarding CSR. In this respect, governments have been pressured by various social agents to adopt initiatives favouring strongly opposed positions. For example, in the European political debate on CSR, governments have come under strong pressure from corporate and business associations not to introduce legislation making CSR policies mandatory. By contrast, civil society in general and NGOs in particular have demanded that government increase its regulation of and control over companies in social and environmental fields.

At a European level, Corporate Social Responsibility is understood as the mechanism for companies to contribute to sustainable development, and specifically to the strategic objectives adopted by member governments and heads of state at the European Council in Lisbon in March 2000, "*by making the European Union the most competitive and dynamic knowledge-based economy in the world, capable of sustainable economic growth with more and better jobs and greater social cohesion.*" In July 2001, the European Commission published a *Green Paper* called *Promoting a European Framework for Corporate Social Responsibility*. This document, coupled with the pioneering action of some member states, such as Denmark and the UK, was a turning point for the legitimization of government action to promote CSR across Europe.

This document was followed in July 2002 by the European Commission Communication: *Corporate Social Responsibility: A business contribution to sustainable development*, advocating the development of national frameworks on CSR. The European Commission asked the governments of EU Members States to take policy initiatives to foster CSR. Governments were asked to: (1) adopt social responsibility policies and preach by example; (2) adopt public CSR policies with regard to employment, social issues, corporate policy, environmental policy, consumer protection, public tenders, foreign and trade policy, and development aid. The European Commission also emphasised the need to disseminate CSR by: pooling experience and best practice; fostering convergence and transparency in CSR practices and instruments; setting up multi-stakeholder forums on CSR.

In 2002, the European Commission promoted the creation of the European Multi-stakeholder Forum on CSR, which produced a series of recommendations in June 2004 from a multi-stakeholder approach of CSR. Nevertheless, in March 2006, the European Commission produced a further communication on CSR, "*Implementing the partnership for growth and jobs: making Europe a pole of excellence on Corporate Social Responsibility*". This latest communication proposes a European Alliance for CSR, as a political umbrella for new or existing CSR initiatives by large companies, SMEs and their stakeholders. In this respect, despite governments' role in fostering CSR, the European Commission decided to give companies a greater role in the field.

This paper aims to contribute to understanding of the government's role promoting Corporate Social Responsibility as an answer to the debate that has arisen within the political and global context during the last decade. This paper: (1) sets out the research findings and the various factors governments or the academic literature consider to be key drivers behind public initiatives for fostering CSR; (2) proposes a relational framework as a model for analysing the various approaches taken by governments and looks at the various public policies that governments may adopt to foster CSR. These policies are classified by the agents involved, and in particular by their relationships; (3) explores the relationship among social, economic and environmental context and the development of CSR governmental approaches; and (4) sets out the governments', companies', and NGOs' perceptions as to what role governments ought to play in fostering CSR.

1. Drivers for Corporate Social Responsibility in the Public Policy Arena

According to a review of the literature produced by governments, international organisations, companies, other stakeholders, and scholars on the subject, there are a series of key driving forces behind governments' promotion of Corporate Social Responsibility. Their role in CSR is seen as part of the search for solutions to the major social, political and environmental challenges facing national and international economies today (Moon 1996, 2004; Zadek et al. 2001; Aaronson & Reeves 2002a, 200b). CSR policy has also been described as an effective way of controlling the negative effects of corporate activity in the context of globalisation, through the promotion of new models of governance (Fox et al. 2002; Crane & Matten 2004; Moon 2004; Albareda et al. 2004; Midttun 2004).

On a national level, the role of governments in CSR has been put forward as both a mechanism to address welfare deficits, and a means of promoting national competitiveness (Zadek et al. 2002). It has also been noted that public policies for CSR are developed in the context of a changing set of exchange relationships between business, government and civil society (Albareda et al. 2004, 2005; Midttun 2004). At both international and national level, CSR policy is developed in the context of an acceptance of CSR as an important contribution to the wider goal of sustainable development (European Commission 2002).

This section outlines some of the key points made in the literature about the drivers for governments to take action for CSR (Table 1).

Table 1

Key drivers highlighted for governments to promote Corporate Social Responsibility

• Globalization and challenges faced by the new economy.
• New models of governance and the welfare state crisis.
• National competitiveness and innovation.
• Sustainable development.
• The partnership approach.

1.1 Globalization and the New Economy

The increasing profile of CSR as a concept in government action is linked to the challenges brought about by globalization and economic change in the late 20th century, such as corporate citizenship, the debate on global governance and the interrelationships between trade, investment and sustainable development (Zadek et al. 2001; Zadek 2001). Globalization has changed the relationships between governments and companies, bringing about an increase in the power of multinational companies over governments, which has been considered an opportunity yet brings with it challenging ethical dilemmas (Crane & Matten 2004).

Some authors (Castells 1996; Zadek et al. 2002) consider national governments have lost the political dominance they once exercised through binding legislation and their role has become more dependent on the influence of multinational companies, who have the freedom to move capital, production processes and goods and services from nation to nation, leaving governments in a vulnerable position.

The legitimacy of governments acting under the influence of business has raised grave concerns regarding ethics and the public accountability of government. In this context, various political innovations and new styles of regulation have evolved, which usually include business, governments and civil society organizations. In the face of these new social challenges, CSR has been described as an important influence on voluntary action by companies and as an instrument for improving business accountability and responsibility

1.2 New Models of Governance and the Welfare State Crisis

In the context of the new globalized economy, political challenges like the crisis in the welfare state and the need to seek new forms of governance in the national context and in the global economy have arisen (Moon & Sochaki 1996; Moon 2004). The welfare state crisis in some countries has made people look for new ways of developing and funding collective action to deal with social demands (Rosdahl 2001) that cannot be met by the state alone such as; poverty; unemployment; lack of economic development and social exclusion of key groups; the crisis of governance and legitimacy.

CSR is seen as a framework within which new ways of collaboration and partnership between business, governments and civil society (Albareda et al. 2004, 2005; Midttun 2005) are used as a mechanism for developing new models of governance to address the major social problems faced by post-industrial societies (Nelson & Zadek 2000; Gribben et al. 2001; Kjaergaard & Westphalen 2001; Wilson 2002). CSR public policies are also considered as a way of promoting good company practices complementing other public efforts for sustainable development.

1.3 National Competitiveness and Innovation

The European Commission (European Commission 2001, 2002) stresses the need to link CSR to the competitiveness of companies and national and regional economies, which in turn has been described as fundamental for a nation's sustainability

Competitive responsibility has been described as the third generation of CSR (Zadek 2001b). Despite the search for a clear link between CSR and the competitive advantage of nations (Zadek & Swift 2002), potential improved financial performance at the micro-level of the company does not necessarily always translate to competitive advantage at the macro level (Zapal 2003). This presents a challenge for governments, which need to find a way of designing and implementing policies that generate leadership and partnership-based innovation, ensuring wide take up of CSR practices across the business community to promote competitiveness at the national level (Tencati 2004).

CSR clusters (Zadek et al. 2003) have been used by governments as a mechanism to promote competitiveness by: helping responsible companies win markets for products; labour and finance; creating frameworks for competition; fostering collaboration or collective response to CSR. Challenge clusters, market-making clusters, partnership clusters and statutory clusters are examples of this CSR clusters help provide a framework for understanding, designing and operationalizing public policies on CSR.

1.4 Sustainable Development

Governments are seen to have a responsibility to promote the economic, social and environmental conditions that favour more sustainable development, and specific commitments on CSR for governments were agreed at the last World Summit for Sustainable Development (Johannesburg, 2002).

The deregulation, downsizing and deficit reduction policies of the 1990's have left governments with the challenge of facilitating the transition to a more efficient, fairer and more sustainable economy (Bell 2005). CSR is widely accepted as the business sector's contribution to sustainable development. CSR is included by some governments within an overarching sustainable development framework, other governments have separate CSR and sustainable development strategies

1.5 The Partnership Approach

The welfare state crisis has made people look for new ways of developing and funding collective action to deal with social dilemmas. The partnership projects that have sprung up with governments, companies and civil society are a response to these needs.

These new alliances have been an important source of innovation at both practical and political levels, bringing about profound structural changes in national and European labour markets, moving from models of collective agreement to one of social partnership between government, business, trade unions and civil society organisations (Nelson & Zadek 2000; Gribben et al. 2001; Kjaergaard & Westphalen 2001; Wilson 2002).

Public–private partnerships (PPPs) are one example of this new model of governance, and have been interpreted as derivatives of the privatisation movement or market compatible forms of stimulating social responsibility (Grimshaw & Vincent 2001). PPPs have also been used to stimulate national competitiveness.

The involvement of civil society organisations as well as governments and business in promoting CSR was clearly legitimised by the European

Commission's creation of a European Multi-stakeholder Forum on CSR, which produced a series of recommendations in June 2004. The European Commission promotes multi-stakeholder dialogue as a form of building consensus on CSR issues by involving all parties concerned (Table 2).

Table 2

Reflections on the literature on governments and Corporate Social Responsibility

An analysis of the literature on the subject leads to the following considerations when analysing government's role in CSR; • CSR public policies cannot be viewed in isolation from the major political and economic problems facing us today. • CSR is a response to and sometimes a result of the new challenges raised by economic globalization. • CSR is a response to the welfare state, CSR isis and presents a new governance model. • CSR is seen as a relational system that links government, company and society initiatives in a joint response to social problems through partnerships. • CSR public policies are linked to national, regional and local economies. • The role of governments in fostering CSR contributes towards sustainable development.

2. The Relational Framework as a Way of Exploring the Changing Role of Government

2.1 Relational Framework

This section presents a relational framework for analysing public policies on CSR. This framework allows us to explore the changing role of government in promoting CSR, adopting a broad perspective that considers the various social agents taking part in the CSR debate. The analytical model attempts to analyse governmental CSR public policies from a relational perspective, employing a CSR policy framework that analyses relationships among the various actors (government, business sector and civil society).

While governments have been described as enablers and drivers of CSR, there has also been increasing recognition of the role of other actors such as civil organizations within the CSR framework. This has called for a new set of relationships between government, business and civil society. It is within the context of this new relational web and the need for a deeper understanding of the role of government in promoting CSR that this research has been undertaken.

In this context governments are now operating in a new relational framework (Mendoza 1996; Midttun 2005), where societal governance is based on a set of increasingly complex and interdependent relationships. The different expectations and perceptions of each exchange relationship need to be addressed to develop public policy for CSR and consideration of these relationships allows a more complete view of government CSR policy.

The relational approach (Mendoza 1996; Midttun 2005) is based on the idea that changes affecting the economic and political structure in recent decades have transformed the roles and capacities of various social agents. This model can be used to give a more complete and holistic view of CSR public policies by focusing on the exchange relationships among these agents. From this perspective, social governance is based on a set of increasingly complex and interdependent relationships. The consideration of these relationships allows a more complete view of government CSR policy that takes in differing expectations, interests, and the challenges of each exchange relationship. Thus, differing expectations and perceptions of each exchange relationship need to be addressed to develop public policy for CSR.

From a relational standpoint, the changing role of government means that governments tend to adopt a multi-stakeholder vision when promoting CSR. (Figure 1)

There are three broad exchange arenas among sectors. Government needs to manage the expectations of theses exchange relationships to facilitate complex inter-organizational networks in which all three sectors play a part. (Figure 2)

Figure 1
The Relational Approach

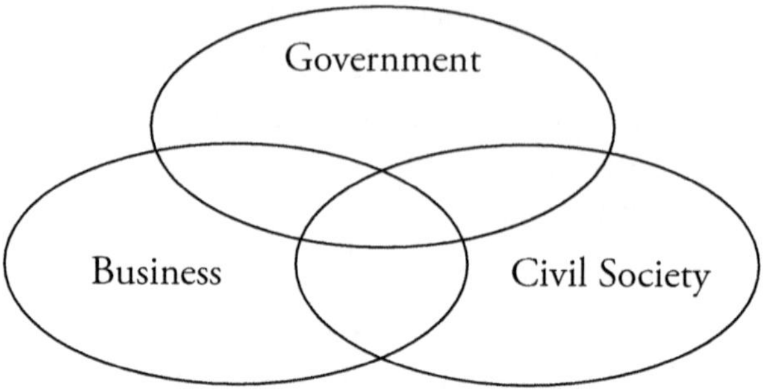

Source: Mendoza, 1996.

Figure 2
Actors and Exchange Arenas

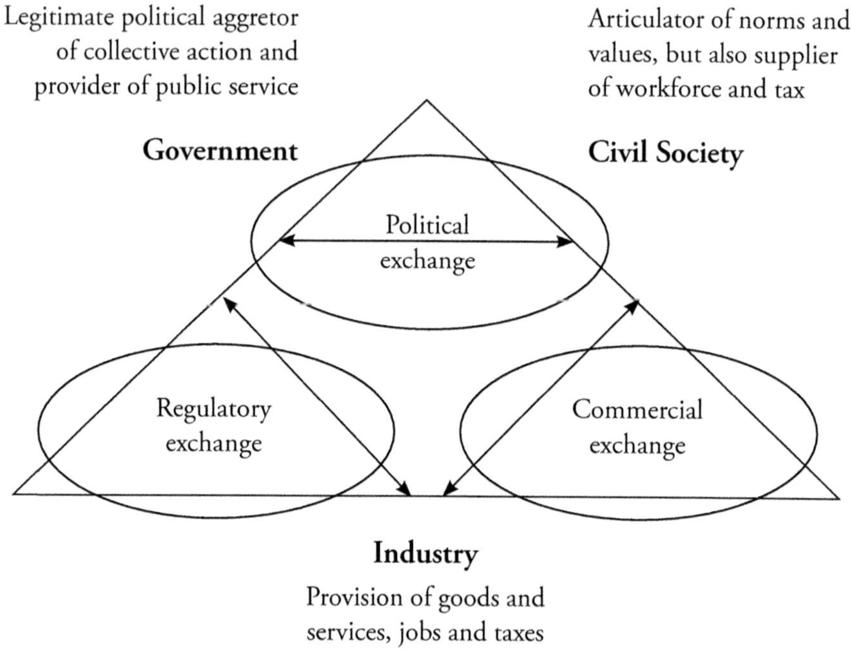

Source: Midttun, 2005.

Applying the relational approach, government policies and programs for CSR can be analysed through the following four relationships (Figure 3):

1. CSR in public administration
2. CSR in government–business relationships
3. CSR in government–civil society relationships
4. CSR in government–business–civil society relationships

Building on the relational state model, the learning from the literature review and other sources (including Albareda et al. 2005) helped us build an analytical framework for grasping government's CSR role. This tool enables analysis of the governmental approach to CSR, from two key perspectives; the overarching policy framework and the policy implementation in terms of specific policies and programs.

Figure 3
Public Policies for CSR in the Relational State Model

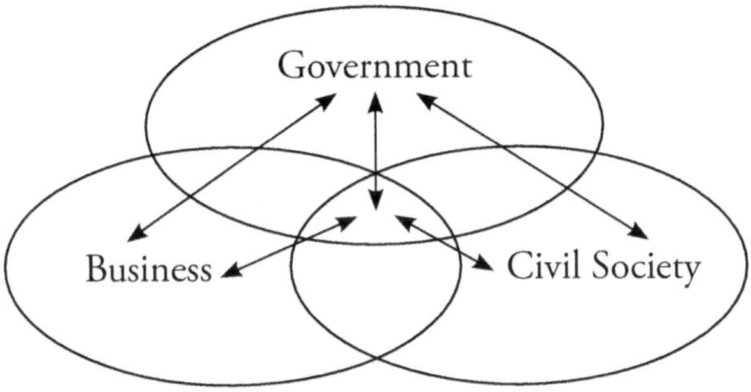

1. CSR in public administration
2. CSR in government–business relationships
3. CSR in government–civil society relationships
4. CSR in government–business–civil society relationships

Source: Albareda, Ysa, Lozano, 2004.

First of all, in order to analyse the changing role of government in CSR, we drew up a table outlining the areas for consideration when analysing the policy framework.

Table 3 highlights CSR vision, objectives, strategy and priorities, governmental structure and policy implementation across different levels of government. Implementation also varies between countries.

Table 3
Government CSR policy framework

Topic	Application
Government CSR policy	Vision
	Objectives, strategies and priorities
Internal government CSR structure	Position of political figure
	Organizational structure
	Centralized or decentralized
CSR responsibilities at different levels of government	CSR cross cutting policies
	Regional/decentralized government
	Local government
Scope of CSR policy	Domestic versus international
CSR role of other organisations	Government agencies
	Intermediary organisations
	Multi-stakeholder organisations
	International organisations

Source: Adapted from Albareda, Ysa and Lozano (2005).

2.2 Policies and Programs

Regarding the second stage, Albareda et al. (2005) developed a broad table of policies and programs that governments may adopt in fostering CSR. Table 3 helps to describe the policy implementation. This list is not intended to be exhaustive or exclusive, additional categories of policies and programs are likely to exist in some countries, and likewise there may be categories which are not relevant in others. This framework can be used to help analyse existing government policies and programs for CSR, compare the situation in different countries or monitor policies and programs over time. These can be categorised according to the sector at which they are targeted, or the sectors involved in their implementation in a relational approach:

- CSR in public administration (i.e. in-house policies);
- CSR in government and business relationships;
- CSR in government and civil society relationships;
- CSR in government, business and civil society relationships.

See Table 4.

Table 4
CSR Policies and Programs

A.1 Government–Public Administration

	Policies	Programs (examples listed here)
INTERNAL ↓	1. Leadership by example	• Action Plan for Government Offices • Work-life balance policies/equal opportunities/ethical investment/ anti-fraud and corruption policies • Accreditation for good employer practices
	2. Creation of internal departments	• Creation of knowledge centres • Creation of monitoring organisations and control systems
	3. Co-ordinating government bodies	• CSR Minister responsible for co-ordinating activities • Cross-government CSR programmes • CSR feasibility studies for new legislation
	4. Capacity building	• Funding for research and innovation programmes • Financial assistance for companies implementing CSR programmes • Publication of guidelines and good practice documents
	5. Public expenditure	• Social and environmental criteria in supplier policies • Ethical purchasing and outsourcing • CSR policies for public contracts
	6. Public campaigns	• Promotion of positive impacts of CSR in business and society • Surveys on public opinion • CSR Awards, communication campaigns and media influence

	International Issues	
	7. International events	• International conferences on CSR • European Commission events • European conferences on CSR
	8. Transferring international debate to local contexts	• Agreements between national and local government • Seminars on geographic or thematic areas • Consideration of CSR regional and local policies
	9. International instruments and agreements	• Promotion of global regulatory frameworks • Development of international certification systems • Creation of evaluation and certification bodies
	10. Foreign trade policy and international development	• CSR integrated into foreign affairs policies for international markets and international development • Promotion of good CSR practice in overseas operations (human rights, labour standards, anti-corruption, environment etc.) • Link CSR to foreign investment policy and international relations
↓ EXTERNAL	Other	

Source: Adapted from Albareda et al. (2005).

A.2 Government–Business

	Policies	Programs (examples listed here)
SOFT ↓	1. Raising awareness	• Identify and promote companies leading in CSR • Promote CSR through websites, publications, specialist journals • Offer CSR services and support to CSR initiatives in companies or partnerships • Undertake surveys and communication campaigns
	2. Voluntary initiatives (facilitating and promoting)	• Promotion of uptake of CSR policies, publication of CSR reports • Encouraging sharing and promotion of good practice • Promotion of SRI, environmental standards, fair trade, sustainable consumption, work life balance, equal opportunities, employee volunteering, employee conditions, lifelong learning • Promotion of business networks • Promotion of public–private partnerships or public–private–civil society partnerships
	3. Capacity building	• Finance research and innovation programmes • Support business–university research programs (instruments, good practice, comparative studies) • Develop guidelines and provide technical assistance • Incentives for sustainability reports
	4. Stakeholders	• Evaluation and communication programs on impact of CSR programs on stakeholders • Market mechanisms to favour CSR (price policies, competition policies, investment principles) • Promote stakeholder dialogue
	5. International	• Incentives for adopting international CSR standards • Promoting CSR good practice in the Third World (labour standards, human rights, anti-corruption)

↓	6. Convergence and transparency	• Promote standardisation across CSR management models, standards, reports, indicators and auditing systems • Promote fair trade labelling systems • Encourage standardization of SRI analysis • Promote inclusion of international CSR agreements in codes of conduct
	7. Evaluation and accountability	• Accountability and auditing mechanisms • Triple bottom line reporting initiatives • Social and environmental labelling
	8. Tax and funding systems	• Tax incentives for CSR (employment Creation, gender balance, work-life balance, environmental initiatives etc.) • Funding streams for CSR (volunteering, social projects etc. • Promote SRI through fiscal mechanisms
	9. Legislation	• Transparency regarding socially responsible investment (pension and investment funds) • Obliging companies to produce sustainability reports • Regulation regarding public contracts and selection processes • Environmental legislation • Adaptation of international agreements to national standards

A.2 Government–Business *(continued)*

	Sector Specific Issues	
	10. SMEs	• Promotion and incentives for good CSR practice in SMEs • Raising awareness of impact of operations in the Third World • Support SMEs in impact assessment • Research on the social and environmental impact of SMEs • Promote the exchange of good practice and business case for CSR in SMEs • Public campaigns directed at SMEs • Encourage cooperation between large companies and SMEs
	11. Community action	• Favourable tax incentives for business in the community • Disseminating good practice and creation of networks
	12. Corporate restructuring	• Work with facilitators to promote CSR in restructuring programmes • Encourage good practice and its dissemination
↓ HARD	Other	

A.3 Government–Civil Society

	Policies	Programs (examples listed here)
SOFT	1. Raising awareness	• Analysis and dissemination of good practice in business operations with high impact on the community (work life balance, social cohesion • Tax incentives for civil society–government partnership programmes • Knowledge dissemination of international agreements with civil society implications (human rights, labour standards)
	2. Voluntary initiatives (facilitating and promoting)	• Campaigns for sustainable consumption, publications, seminars and dissemination • Ethical investment initiatives • Support SRI initiatives • Support socially responsible consumption
	3. Capacity building	• Publications, events, press • Surveys and CSR Awards
	4. Stakeholders	• Create communication mechanisms to foster business–community dialogue • Promote transparancy mechanisms • Promote partnerships and participate in them
	5. International	• Promote initiatives with international NGOs • Participation in international civil society activities
	6. Convergence and transparency	• Fair trade labelling schemes • Social enterprise definitions
	7. Evaluation and accountability	• Accountability and auditing mechanisms • Triple bottom line reporting initiatives • Social and environmental labelling
↓ HARD	8. Tax incentives and funding streams	• Support for government–civil society partnerships for CSR initiatives
	Sector Specific Issues	
	9. Management of social organizations	• Improve management of social enterprise • Enable SRI • Environmental and social criteria for public contracts to socially financed organisations
	Other	

Source: Adapted from Albareda et al. (2005).

A.4 Government–Business–Civil Society

	Policies	Programs (examples listed here)
SOFT ↓	1. Raising awareness	• Sharing good practice and knowledge dissemination • Create National Resource Centres (e.g. National Contact Point, The Netherlands)
	2. Capacity building	• Multi-stakeholder Forums • Business support networks • Sharing experience and best practice
	3. Voluntary initiatives (facilitating and promoting)	• Round tables on codes of conduct • University–business research projects, promoting dialogue • Proactive role in promoting innovation, pilot projects, dialogue
	4. Stakeholders	• Consumers; information on supply chain, sustainability index of products • Investors: information on CSR policies and expectations regarding pensions
	5. Evaluation and accountability	• Accountability and auditing mechanisms • Triple bottom line reporting initiatives • Social and environmental labelling
	6. Convergence and transparency	• Management standards • Codes of conduct • Promote simple and flexible indicators
HARD	7. International	• International partnerships • Networks and alliances • Multi-stakeholder forums

	Sector Specific Issues	
	8. Community action	• Urban regeneration projects • Education projects in poor areas
	9. Muli-sectoral partnerships	• Promoting CSR networks with public/private participation • New social partnerships and common frameworks • Local partnerships between different actors for urban regeneration • Stakeholder involvement in developing guidelines • Bringing together different sectors
	10. Socially responsible investment and fair trade	• Pension schemes with social, environmental criteria • Transparency in definition of SRI • Selection, retention and realization of investment with CSR considerations • Consumer rights
	Other	

Source: Adapted from Albareda et al. (2005).

3. Exploring the Relationship among Social, Economic and Environmental Context and the Development of Corporate Social Responsibility Governmental Approaches

3.1 Social, Economic and Environmental Context

This section presents the first working hypothesis, which was based on the literature review and the relational framework and whose purpose was to analyse the CSR policies of European governments. The literature review revealed that there are various challenges regarding the future of the welfare state (employment, ageing population, social cohesion and inequality, competitiveness) and sustainable development in each country. This item linked the hypothesis on the relationship between the social, economic and environmental contexts and the development of CSR governmental approaches in each country or region.

An initial hypothesis suggests that the CSR issues on the agenda of each government are related to the specific deficits in the economic, social and environmental systems faced in a given country. We noted that the structure of CSR policies in some countries is directly linked to the main socio-economic challenges facing the nation. In Italy, for example, CSR policies are linked to innovation, competitiveness and social cohesion; in the UK, to policies on social cohesion, unemployment and environmental sustainability; in Norway, to international competitiveness, international trade and environmental pressures; in France, to policies on sustainable development and worker participation; in Spain to social issues, social inclusion and reconciling family life with work.

In future research, analysis of the socio-economic and environmental indicators should contribute to a better understanding of the specific challenges that governments may address through their CSR policies.

The key performance indicators outlined in Table 5 have been identified as being critical from the standpoint of government CSR policies to help describe a country's socio-economic and environmental context. Most of these indicators were selected to assess the implementation of the EU Sustainable Development Strategy.

Table 5

Performance indicators to analyse the specific drivers of government CSR policy

• *Economic growth* • *Employment* • *Inequality in income distribution* • *Population ageing* • *Competitiveness/innovation* • *Environmental impact*
Economic growth • Fundamental dimension of development to be taken into account in assessing the wealth-Creating capacity and competitiveness of a given nation. • Indicator: Real GDP growth rate (growth rate of Gross Domestic Product at constant prices – 1995 – percentage change on previous year)

Employment • Slowing growth in total employment and high unemployment rates are two of the most important drivers of active social policies by governments and of the building of public–private partnerships. • Indicator: The total unemployment rate (i.e. the unemployed as a percentage of the labour force).
Inequality in income distribution • Public efforts against social exclusion for more inclusive paths of development are crucial elements for valuable government welfare policies, of which CSR programmesprograms can be an important tool in a welfare mix perspective. • Indicator: The Gini index, the most widely used tool in this field, which measures the extent to which the distribution of income (or consumption) among individuals or households within a country deviates from a perfectly equal distribution. A value of 0 represents perfect equality, a value of 100 perfect inequality.
Population ageing • The growing weight of elderly persons in the Western societies constitutes a major challenge for governments and calls for an active role of the private (for profit and not for profit) sector in related policies (public health, pension management, assistance etc.). • Indicator: Eurostat current and projected old age dependency ratio between the total number of elderly persons of an age when they are generally economically inactive (aged 65 and over or aged 60 and over, depending on the context) and the number of persons of working age (from 15 to 64 or from 20 to 59, depending on the context).
Competitiveness and innovation • Considering sustainability, and innovation to address the challenges it presents, as a key-factor of business success, the monitoring of R&D expenditures provides important insights on corporate capacity to face competition. • Indicator: the gross domestic expenditure on research and development as a percentage of GDP.
Environmental impact • The Ecological Footprint is a way of measuring a country's natural resources consumption and can be compared with nature's ability to renew these resources. In 2001, humanity's Ecological Footprint was 2.5 times larger than in 1961, and exceeded the Earth's biological capacity by about 20 percent. • Indicator: Total area required to produce the food and fibre that a country consumes, absorb the waste from its energy consumption, and provide space for its infrastructure.

Source: Tencati 2005.

4. The Challenges Faced by the Various Stakeholders and their Perceptions

This section analyses an implicit aspect of governmental CSR policies which, despite its importance, does not appear in government statements. We highlight some of the perceptions from various stakeholders of the key issues to consider when comparing, monitoring or creating public policy for CSR. The results of this section are based on interviews carried out with a small sample of representatives of businesses, government and civil society in Italy, Norway and the UK. The comments in this section are therefore not representative of each sector, however similarities have been found in the perceptions of people working in the same sector in different countries.

As noted previously, the practices of different governments vary greatly. However, there are common trends across countries regarding what various sectors think government's role should be. This section highlights examples of the different perceptions from each group, as well as the perceptions that are common across the three countries.

4.1 The Role of Government in Corporate Social Responsibility

Governmental perceptions

- CSR is considered a cross-governmental issue, with a broad agenda related to social, environmental and international issues. Different ministries introduce CSR elements into their specific policy areas in parallel with each other, however there is often limited co-ordination between these initiatives.

- There is a strong link between CSR and sustainable development, in the UK and in Italy CSR is seen as the business contribution to this agenda, the Norwegian Ministry of the Environment is introducing CSR elements into their sustainability agenda.

- CSR is generally seen as a strategic and competitive opportunity, however how this is structured in government varies greatly.

- A multi-stakeholder partnership is seen as appropriate for CSR.

Business perceptions

- Governments should adopt a leading role, define frameworks of action to influence and encourage other organisations (businesses and different levels of government).
- The relationship with CSR and sustainable development policy is unclear.
- Governments should be consistent in their policies, both in their own practice and through promoting an advanced CSR agenda at international level.
- CSR policy does not always reach business. In the UK, there appears to be more successful engagement with the NGO community. In Norway, the SME community has strong CSR practices that take little account of government CSR policy.

Civil Society perceptions

- CSR became a key issue after the liberalisation policies of the 1980s, and following the decline in the provision of state services it has been seen as a mechanism to provide innovative solutions to the need to renew traditional welfare state mechanisms.
- The most important role for government is to act as broker between sectors, working with both the supply side of CSR (companies, consultants, industrial associations) and the demand side of CSR (citizens, consumers, investors, stakeholder groups).
- Government should reward good practice, e.g. supporting social enterprises whose activities benefit the community.
- There have been good attempts by the UK government to promote CSR in companies operating overseas, however more should be done with the SME sector.

See Table 6.

Table 6

Common perceptions on CSR policy	
Government	• Should be largely business driven • Should be cross-sector • Cross-government • Business contribution to sustainable development • A strategic and competitive opportunity • Multi-stakeholder and partnership approach is best
Business	• Government has an important role in CSR • Government should influence all social agents, companies and civil society organizations • Government should define a clear policy framework and lead by example
Civil society	• No common perception identified, varying opinions

4.2 The Legislative versus Voluntary Perspective

Governmental perceptions

- A light touch, voluntary approach is taken in Italy and the UK as CSR is seen as more effectively implemented if it is business led and not regulated. In both countries there is a general preference for a partnership approach with business, raising awareness, capacity building, stakeholder engagement, and facilitation of voluntary initiatives.
- In Norway however with a strong welfare state tradition, there is still commitment to legislative methods and therefore little scope for voluntary action on a domestic level. There is potential for the promotion of CSR on a voluntary basis abroad.

Business perceptions

- Clearly understandable and simple regulation is essential, with a well defined level playing field with basic rules for all players. Business should be demanding this and at least complying with minimum standards.

- Governments should continue with their facilitating role, encouraging business to behave more responsibly, however they should not enforce standards until a certain amount of experimentation has been undertaken and good practice been developed. This way, leading companies should identify future CSR standards to encourage other companies to follow suit, before governments impose these standards.
- Government should encourage innovation, and facilitating partnerships among well-run firms is one way to achieve this. However innovation for sustainability will ultimately rest in the hands of business.
- In Norway there is general endorsement of the international CSR focus, although there is a call for more policy support for SMEs in the domestic market.

Civil Society perceptions

- Legislation has its limitations in terms of promoting CSR and governments should adopt a role of encouragement and support, as well, as this is more likely to lead to compliance.
- The "ideological" debate on voluntarism can be interpreted as an excuse to avoid legislation which may be politically difficult to impose.
- In the case of Norway, trade unions are keen to solve domestic issues by traditional bi- or tripartite negotiations and NGOs (which support a stronger domestic CSR policy focus and tough legislation. This is particularly true of environmental NGOs).

See Table 7.

Table 7

Common perceptions on legislative versus voluntary perspective	
Government	• Preference for voluntary approach
Business	• Governments should facilitate not overdo legislation • More policy support needed and level playing fields
Civil society	• Support for approach based on partnership and engagement • Varying opinion on voluntary approach

4.3 *CSR Policy and Sustainable Development*

Governmental perceptions

- There is agreement that CSR is not only the business contribution to sustainable development, but that it is a crucial part of the success of this wider goal.
- Norway has a pioneering position to defend in terms of sustainable development, however it is currently facing serious dilemmas with its large petroleum industry.
- In the UK, the links between CSR and sustainable development have not been strong enough, or clearly communicated and in the case of Norway the merging of these two agendas came fairly late.

Business perceptions

- Sustainable development and CSR are about the same issues, and CSR is useful as a broad general term about how business needs to do more to respond to the outside world.
- In Norway, a recent survey shows business expects CSR and sustainability to be important in the future and there is general support for government's engagement in multi-lateral initiatives for sustainability and CSR.
- The major future challenges for government and economic players are: CO2 reduction; waste prevention; reduction and recovery; and renewable energies.

Civil Society perceptions

- NGOs have widespread expectations of improving the social and environmental performance of companies, for example in Norway there is extensive concern over Norwegian petroleum prospection and production in the Arctic.
- In Italy CSR is seen as fostering the sustainable development of companies by increasing strategic resources (e.g. qualified employees, stronger reputation and broader social consensus).

- There is broad consensus that the two agendas cover the same ground, and clear links are made with the Lisbon Strategy, the sustainability of the EU and CSR.
- In the UK there is a call for better co-ordination between the two strategies.

See Table 8.

Table 8

Common perceptions on CSR policy and sustainable development	
Governments	• CSR is described as the business contribution to sustainable development
Business	• CSR and sustainable development are major challenges for both governments and businesses of the future
Civil society	• There are strong links between both agendas • High expectations of business practice and how governments incorporate the agendas into a policy framework

4.4 CSR Policy and Competitiveness

Governmental perceptions

- CSR is seen as a strategic and competitive opportunity for companies, in Italy, particularly for SMEs and in Norway particularly for industry abroad.
- CSR enhances reputation and can stimulate competitors to work in the same area.
- However, in Norway domestic policy-oriented Ministries tend to prefer more traditional legislative tools and do not set any store by CSR and competitiveness, the Ministry of Finance takes a cautious approach and promotes a moderate SRI policy for the Petroleum Fund.

Business perceptions

- There is some scepticism concerning this relationship although some acceptance of the potential for a win–win situation.
- Possible business benefits include improved quality in the processes and products of a company, greater self-knowledge internally and externally, allowing quick response to changing markets, and increased reputation.
- In the case of Norway international business welcomes the CSR support from government, as it helps build the reputation as a responsible business, in line with domestic welfare state norms.

Civil society perceptions

- There is a clear link between the Lisbon Strategy, sustainability and CSR in the European economy and – if CSR becomes a crucial variable – it could become a key factor for the competitive success of national economies.
- One of the objectives of government in promoting CSR is to stimulate innovation and competitiveness. Governments fund partnerships between NGOs and business in the belief that they will improve a region's competitiveness.
- Partnerships between business and environmental NGOs in Norway combine green innovation with lobbying, and there is a strong focus on international issues such as human rights and poverty.
- While CSR is recognised as important for corporate success, defining new strategic ways of doing business and the creation of value that is both pro-active and respects different constituencies can have a great impact at the macro level.

See Table 9.

Table 9

Common perceptions on CSR policy and competitiveness	
Government	• CSR is a strategic and competitive opportunity for companies
Business	• Support CSR policies that focus on competitiveness and innovation • CSR could be part of an innovative development path
Civil Society	• Governments promote CSR to stimulate innovation and competitiveness • CSR could become the key factor for the success of national economies in global markets

4.5 *CSR Policy and the Welfare State*

Governmental perceptions

- There is a relationship between these two areas, however it is described in different ways. In the UK, CSR policy is defined as "part of modern welfare state policies"; in Italy, the relationships between CSR and the welfare policy are considered as constituting a new welfare mix; in Norway, the two areas "run in parallel".
- The case of Norway is clearly different from Italy and the UK, where the present government has reinforced its welfare state model, the CSR agenda domestically is developed in this context.

Business perceptions

- In Italy, where companies have a fundamental social role, they are seen as the ones who should provide innovative solutions to the needs of civil society as expressed through market forces.
- In the UK, the lead comes from government, which enlists corporate help in the form of privatisation of public services in areas where it lacks the resources to meet its obligations. However, companies are only willing to make a contribution if it makes business sense. Philanthropic reasons are not enough.

- In Norway, the welfare state is largely taken as given, and business relates to this, focusing more on CSR issues abroad, although CSR issues are beginning to feature as supplementary elements in public–private partnerships.

Civil Society perceptions

- In Italy and the UK there is an acceptance that traditional welfare state mechanisms need to be renewed and that business has a role in this, however how government relates to business on this issue differs across countries.
- In the UK, where businesses are seen to contribute to government priorities more than in other European countries, government exerts considerable pressure through its policies (e.g. unemployment initiatives, New Deal program). This can lead to confusion over business' role, contributing to public agendas (e.g. sponsoring training programs for young people) or taking over formerly public services such as education (e.g. privately run schools).
- Business reacts badly to multiple initiatives from government encouraging business participation in social issues, and the perception that CSR is being used to secure additional funding for areas with depleted budgets.
- In Norway trade unions insist on formal bilateral negotiations or legislation in social areas, rather than soft voluntary measures, however NGOs with weaker neocorporatist traditions, are moving beyond welfare state policies and more willing to be involved in a CSR mode.
- In Italy, the "Civil Economy" used to describe social enterprises whose activities benefit the community, should be supported and protected by government. CSR can provide innovative solutions to address the need to renew traditional welfare state mechanisms through encouraging the active involvement of companies in social issues, through voluntary and strategy-driven measures.

See Table 10.

Table 10

Common perceptions on CSR and the welfare state	
Government	• The relationship between these areas depends on the welfare state background of the country
Business	• The relationship between CSR and welfare state is demonstrated through the governments' public partnership strategies
Civil Society	• CSR can provide innovative solutions to address the need to renew traditional welfare state mechanisms, looking for companies with socially responsible policies

4.6 Exchange Relations between Sectors

Governmental perceptions

- The relationships between NGOs, business and government are considered fundamental for sustainable development, and in the case of Italy for a more cohesive and inclusive society.
- Government adopts an impartial role, facilitating the participation of all sectors when drafting policy.
- Government has to ensure CSR policy fits the business agenda, as well as taking into account the concerns of civil society, and in particular NGOs. In the case of the UK, the DTI leads CSR policy and its primary concern is business development.
- Businesses and NGOs have a strong idea of each other's motivations, whereas government sits in the middle of the two.

Business perceptions

- Government should adopt the role of mediator, encouraging good practice and encouraging business to provide the solutions to society's needs.
- Government should work with civil society to create the right framework to achieve change through market forces.

Civil Society perceptions

- The relationships between business, government and civil society have changed a great deal since 1992 and are now much more dynamic.
- NGOs are important players in the debate, and have a powerful influence over business practice, in some cases more powerful than that of government. This is largely assisted by the media and the internet.
- The activist NGO movement has played a critical role in controlling business, however other members of civil society have adopted the role of partners with industry or as experts in their fields.
- Trade unions negotiate behind closed rooms with business and government whereas NGOs such as Greenpeace and Amnesty International, WWF and Transparency International thrive in the media limelight.
- The attention of NGOs on business practice has increased global scrutiny and accountability of corporate behavior, however there is a debate about the legitimacy of NGOs as representatives of civil society and their own accountability towards society.
- A brokerage function between the three sectors is needed, and this may be best led by business as government-led initiatives are not always the most successful.
- In the case of Italy, there is a sense that government needs to strengthen its engagement with CSR with civil society, as the focus has so far largely been on industry.

See Table 11.

Table 11

Common perceptions on the relational state	
Government	• CSR gives a new dimension to the relationships between government, business and civil society
Business	• Government should act as mediator between business and civil society to provide solutions to society's needs expressed though markets • Government needs to create level playing fields to allow multi-stakeholder dialogue
Civil Society	• Civil society exercises great influence, particularly through the media, over both governments and business • Different organisations adopt different roles, some work in partnership, others mobilize action through confrontation • CSR requires multi-stakeholder dialogue and partnership strategies

4.7 Overall Perceptions from Each Sector

The perceptions of governments and businesses are similar across the three countries and generally support government CSR policies. The perceptions of NGOs are more varied.

Governmental perceptions

- CSR is an issue that has been incorporated into the government's agenda and is the business contribution to sustainable development.
- CSR has to be a business-driven approach, even though government has a role to play.
- CSR is a cross-governmental and cross-sectoral policy.
- CSR is a strategic and competitive opportunity for national and regional economies.
- A systematic approach and a co-ordinating national framework is needed.

- The lack of consistent CSR public policy (with sustainable development, with government's own operations, in different countries) generates confusion and absence of clarity for companies and stakeholders
- In Italy and the UK, CSR should be based on voluntary initiatives domestically and abroad, in Norway there is a strong commitment in domestic policy to traditional legislative methods.
- For CSR the relationships between NGOs, business and the government are fundamental and the role of government in relation to other stakeholders needs to be clarified.

Business perceptions

- Governments have a key role to play in CSR.
- Government should play a role of facilitator, but must not overdo legislation. They should be encouraging businesses in a certain direction, and only legislate after allowing a period of experimentation and innovation in this area.
- Governments should define a level playing field with basic rules for all players.
- Governments have to represent many interests and often adopt an impartial role, but they should be leaders in influencing other social agents.
- Governments should be consistent between their CSR policies and their own practices.
- Central government should define a clear, sound framework for local authorities and companies.
- Governments should facilitate an innovative and competitive framework for companies adopting CSR policies domestically and abroad. In some cases, there is scepticism about the relationship between CSR and competitiveness.
- Government needs to adopt the role of mediator, working with civil society and business to achieve a paradigm shift, through market forces.

Civil Society perceptions

- There are different views on the role that government should adopt. Some consider that governments should only provide encouragement and support, while others argue that government should perform a more regulatory role.
- Most of the NGOS interviewed support the voluntary approach to CSR but others disagree.
- Government should better co-ordinate their sustainable development strategy and CSR policies.
- There is a link between CSR policy and innovation and competitiveness.
- CSR is important for corporate success, contributes to a nation's competitiveness and supports different paths of local development.
- CSR can provide innovative solutions to address the need to renew traditional welfare state mechanisms through encouraging a more active role by companies.

5. Final Thoughts and Key Issues to Consider

Having analysed these four sections, we present the fundamental elements that enable one to analyse the role of governments in promoting CSR public policies. We highlight some of the elements revealed by this research (which is of an exploratory nature given that not all countries and regions were intensively analysed). This chapter thus provides an initial approach to the role of governments in fostering CSR. This approach will need to be broadened theoretically and analytically in future research.

Basically, the objective of this research was to explore a policy framework in order to analyse the role of government in promoting CSR in certain European countries. The use of the policy framework helps gain a broader view of the governmental role in fostering CSR.

First, analysis of the literature has helped us grasp the various drivers behind government policies promoting CSR. These drivers are the general challenges facing national governments within a new context of economic, political, social and cultural globalisation. As a result, governments need to develop new models of social governance that involve other social agents such as companies and civil society. Other aspects linked by the literature to CSR is the crisis in the Welfare State and the need to come up with initiatives that improve competitiveness and national innovation in global markets, and the challenge of achieving sustainable development in the context of social and environmental limits to growth.

Second, we have seen how the boundaries of business' role in society have become blurred over the last decade in certain fields. Government may encourage the involvement of business in fields where public services are lacking. However, boundaries can also blur as some businesses provide public services while others manage their relationships with stakeholders in a way that benefits the community. In order to respond to this new situation, today's governments need to manage a complex set of relationships between sectors and among social agents.

Third, we subscribe to the idea that all these aspects can be linked to the general consensus on a multi-stakeholder approach to CSR public policy advocated by the European Commission and many governments in Europe and elsewhere. Although social capital and the traditional role of business in society is an important factor, most agents agree that a partnership approach is the most effective approach to CSR public policy implementation. In this context, government has been seen as a broker among agents. It is generally accepted that government should assume the role of mediator, creating a common framework among social agents.

Bearing in mind the foregoing aspects, the application of the relational framework set out in the second section of this paper facilitates exploration of specific public CSR policies and takes into account the relationships among governments and those the policies are aimed at. An initial analysis of this relational framework regarding public CSR policies in three European countries (Italy, Norway, and the United Kingdom) indicates that the key issues to consider are indicated in Figure 4.

Figure 4

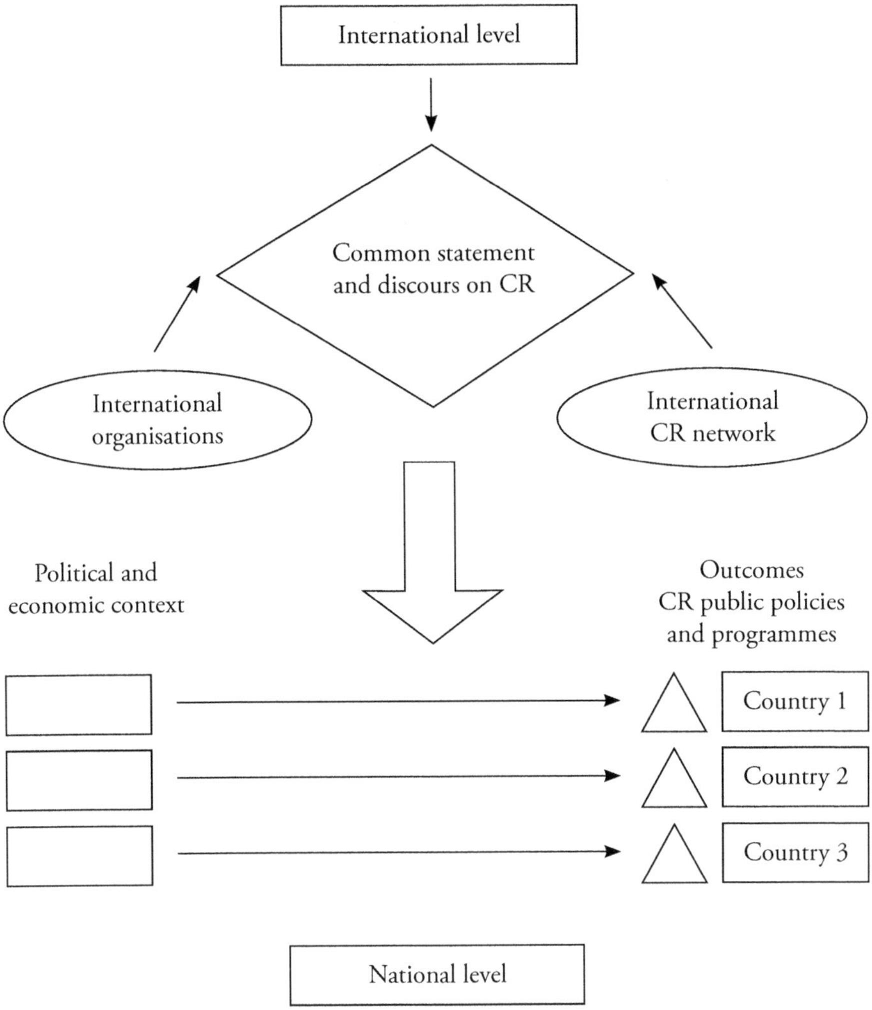

1. At the international level, there is a common discourse on CSR, shared by various international organisations such as the European Union, the UN Global Compact, or the International Labour Organization and the international CSR network (CSR Europe, World Business Council for Sustainable Development, Business for Social Responsibility, Forum Empresa among others). However, the application of

this discourse into specific governmental polices and programs varies across countries. In general, Governments have a broader common perception of the concept of CSR. Nevertheless, governments show differences in applying this concept to political frameworks and implementation strategies. In general, different governments choose different yardsticks in their conceptualisation of CSR. These yardsticks are provided by various organisations: the European Union; the UN (Global Compact); the business contribution to sustainable development; or in some cases, business' role in addressing the problems faced by society today. However, governmental policy frameworks and policy implementation vary regarding: their vision, objectives and priorities; their scope (international versus domestic); organizational structures; initiatives; policy mechanisms and programs; levels of implementation at regional and local levels.

2. As we have seen, CSR policies are directly linked to the globalised context in which governments find themselves; the increasing power of companies over governments; and the crisis in the Welfare State. In parallel with this, we noted how CSR policies can be indirectly linked to social, economic, and environmental challenges facing a given country. In this respect, we consider that the links need to be analysed between public CSR policies and social, economic and environmental indicators of economic growth, unemployment, inequality in income distribution, population ageing, competitiveness/innovation and environment in each country. Such links ought to be explored in future research.

3. In many countries, CSR public policies focus mainly on relations between governments and companies, although in some cases they cover policies addressing the relationships between government, companies, and civil society. If CSR is defined by its multi-stakeholder and partnership characteristics, this relational aspect needs to be a basic aim of many governmental policies and programs. Our research findings indicate that the relational analysis method provides an approach to specific governmental policies that takes into account both governments' vision and strategies, and the social agents at whom the strategies are aimed.

Accordingly, this research should be broadened to provide specific analysis of public CSR policies adopted by all countries. In Europe, a more precise profile is needed for countries by geographic area (English-speaking Central and Eastern European, Scandinavian, Mediterranean, and other continental countries). In the case of Eastern Europe, the transformation from communist to market-based economies contributes a distinct "ideal type" to the European political economy landscape. Further differentiation in Central and East European traditions would make it advisable to eventually include several cases in the broader European comparison.

Nevertheless, the research should also provide data permitting comparison of European countries with other areas of the world. Global comparisons need to be drawn between developed and industrialized countries on the one hand, and developing and underdeveloped nations on the other. This broader research will give us a global map of the role of government in promoting CSR public policies and the multi-stakeholder approach to CSR, allowing us to analyze the relationship between public CSR policies in a global context. This will facilitate analysis of complex relationships between social agents, revealing a new role for governments as promoters and catalysts for socially, environmentally responsible policies in partnership with social actors.

References

Aaronson, S. & Reeves, J. (2002a). *The European Response to Public Demands for Global Corporate Social Responsibility.* Washington, D.C.: National Policy Association.

Aaronson, S. & Reeves, J. (2002b). *Corporate Social Responsibility in the Global Village: The Role of Public Policy.* Washington, D.C.: National Policy Association.

AccountAbility; Business for Social Responsibility (2003): *Business and Economic Development: The Impact of Corporate Social Responsibility Standards and Practices.* AccountAbility.

Albareda, L., Ysa, T., Lozano, J. M. (2004). "The Role of Public Policies in Promoting CSR: A Comparison among the EU–15." Paper presented at the Interdisciplinary CSR Research Conference, Nottingham.

Albareda, L., Ysa,T., Lozano, J. M. (2005). The Role of Governments in Fostering CSR. In A. Kakabadse & M. Morsing (Eds), *Corporate Social Responsibility. Reconciling Aspiration with Application*. Houndmills: Palgrave.

Beck, U. (1986). *Risikogesellschaft. Auf dem Weg in eine andere Moderne*. Frankfurt am Main: Suhrkamp.

Bell, D. V. J. (2005). *The Role of Governance in Advancing Corporate Sustainability*. Background Paper. Final Draft. York University, Canada: Sustainable Enterprise Academy.

Benbeniste, S., Pueyo, R., Llaría, J. (2005). *Responsabilidad social corporativa y políticas públicas. Informe 2004*. Zaragoza: Fundación Ecología y Desarrollo.

Bendell, J., & Kearins, K. (2004). "Remembering the Politics: The Emergent Political Dimension to Corporate Social Responsibility and its Management." Paper presented at the Interdisciplinary CSR Research Conference. University of Nottingham, ICCSR, 22–23 October.

Castells, M. (1996). *The Information Age: Economy, Society and Culture*, Vol. 3. Cambridge, MA: Blackwell Publishers.

CBSR (2001). *Government and Corporate Social Responsibility. An Overview of Selected Canadian, European and International Practices*. Vancouver: Canadian Business for Social Responsibility.

Chatam House, Sustainable Development Programme (2004). *Following up the World Summit on Sustainable Development Commitments on Corporate Social Responsibility: Options for Action by Governments*. Executive Summary. London.

Crane, A. & Matten, D. (2004). *Business Ethics. A European Perspective. Managing Corporate Citizenship and Sustainability in the Age of Globalization*. Oxford: Oxford University Press.

Donaldson, T. (2001). The Ethical Wealth of Nations, *Journal of Business Ethics*, Vol. 31, pp. 25–36.

DTI (2001). *Business and Society Developing Corporate Social Responsibility in the UK*. London: UK Government, Department of Trade and Industry.

DTI (2003a). *Business and Society. Corporate Social Responsibility Report 2002*. London: UK Government, Department of Trade and Industry.

DTI (2003b). *Corporate Social Responsibility. A Draft International Strategic Framework*. London: UK Government, Department of Trade and Industry.

DTI (2004). *Corporate Social Responsibility. A Government Update.* London: UK Government, Department of Trade and Industry.

Elkington, J. (1998). *Cannibals with Forks. The Triple Bottom Line of 21st Century Business.* Toronto: New Society Publishers.

Esping-Andersen, G. (1999). *Social Foundations of Post-Industrial Economies.* Oxford: Oxford University Press.

European Commission (2001). *Promoting a European Framework for Corporate Social Responsibility.* Green Paper, COM (2001) 366-final. Brussels.

European Commission (2002). *Corporate Social Responsibility: A Business Contribution to Sustainable Development.* Communication, COM (2002) 347-final. Brussels.

European Commission (2003). *The Role of Public Policies in Promoting CSR.* Conference Booklet, EU Presidency Conference, Venice, 14 November.

European Commission (2004). *Corporate Social Responsibility. National Public Policies in the European Union.* Brussels.

European Multistakeholder Forum on CSR (2004). *Final Results & Recommendations.* Brussels, 29 June.

Eurostat (2005). *Sustainable Development Indicators.* http://epp.eurostat.cec.eu.int, site visited: 2–5 June 2005.

Fox, T., Ward, H., Howard, B. (2002). *Public Sector Roles in Strengthening Corporate Social Responsibility: A Baseline Study.* Washington, D.C.: The World Bank.

The Frank Hawkins Kenan Institute of Private Enterprise (2003). *Promoting Global Corporate Social Responsibility. The Kenan Institute Study Group Consensus.* Washington: The Frank Hawkins Kenan Institute of Private Enterprise. Policy Recommendations.

Gribben, C., Pinnington, K. Wilson, A. (2001). *Governments as Partners: The Role of the Central Government in Developing New Social Partnerships.* Copenhagen: Copenhagen Centre.

Grimshaw, D. & Vincent, S. (2001). New Control Modes and Emergent Organisational Forms: Private–Public Contracting in Public Administration. *Administrative Theory & Praxis,* Vol. 23, No. 3.

Guarini, E. & Nidasio, C. (2003). "CSR Role in Public–Private Partnerships: Models of Governance." Paper presented at the research track 6: "The Policy Framework/The Societal Context" in the 3rd Annual Colloquium of the European Academy of Business in Society. Copenhagen.

Habisch, A., Jonker, J., Wegner, M., Schmidpeter, R. (Eds) (2005). *Corporate Social Responsibility Across Europe.* Berlin: Springer.

Italian Ministry of Labour and Social Affairs (2003). *Project CSR–SC: The Italian Contribution to CSR Promoting Campaign Developed at European Level.* Rome.

Jespersen, K. (1994). Our Common Concern. Conference. Oslo.

Joseph, E. (2003). *A New Business Agenda for Government.* London: Institute for Public Policy Research.

Kjaergaard, C. & Westphalen, S. (2001). *From Collective Bargaining to Social Partnerships: New Roles of the Social Partners in Europe.* Copenhagen: The Copenhagen Centre.

Lafuente, A., Viñuales, V., Pueyo, R., Llaría, J. (2003). *Responsabilidad social corporativa y políticas públicas.* Madrid: Fundación Alternativas.

Lepoutre, J., Dentchev, N., Heene, A. (2004). "On the Role of the Government in the Corporate Social Responsibility Debate." Paper presented at the research track 7: "Policy Making and the Role of Government" in the 3rd Annual Colloquium of the European Academy of Business in Society: "The Challenge of Sustainable Growth: Integrating Societal Expectations in Business." Brussels.

Loh, J. & Wackernagel, M. (2004). *Living Planet Report 2004.* Ghent: WWF.

Lozano, J. M., Albareda, L., Ysa, T., Roscher, H., Marcuccio, M. (2005). *Los gobiernos y la responsabilidad social de las empresas. Políticas públicas, más allá de la regulación y la voluntariedad.* Barcelona: Granica.

Mendoza, X. (1991). *Algunas reflexiones acerca de la 'transición al mercado' de los servicios sociales.* Barcelona: Jornades Públic–Privat i Benestar Social.

Mendoza, X. (1996). Las transformaciones del sector público en las sociedades avanzadas. Del estado del bienestar al estado relacional, *Papers de Formació,* No. 23. Diputació de Barcelona.

Midttun, A. (2003). "Business, the State and Civil Society: Introduction to a CSR Research Programme." Paper presented at the research track "The policy framework/societal context" in the 2nd Annual Colloquium of the European Academy for Business in Society.

Midttun, A. (2004). "Realigning Business, Government and Civil Society: The C(S)R Model Compared to the (Neo)Liberal and Welfare State Models." Paper presented at the research track 7: "Policy Making and the Role of Government" of the III Colloquium of the European Academy of Business in Society "The Challenge of Sustainable Growth: Integrating Societal Expectations in Business." Brussels.

Midttun, A. (2005). Policy Making and the Role of Government Realigning Business, Government and Civil Society. *Corporate Governance. The International Journal*, Vol. 13, Issue 4.

Moon, J. & Richardson, J. J. (1984). The Unemployment Industry. *Policy and Politics*, No. 12, pp. 391–411.

Moon, J. & Sochaki, R. (1996). The Social Responsibility and New Governance. *Government and Opposition*, No. 27, pp. 384–408.

Moon, J. (2002). Business Social Responsibility and New Governance. *Government and Opposition*, No. 37, pp. 385–408.

Moon, J. (2004). Government as a Driver of Corporate Social Responsibility: The UK in Comparative Perspective. *ICCSR Research Paper Series*, No. 20, pp. 1–27. The University of Nottingham.

Moore, C., Richardson, J. J., Moon, J. (1985). New Partnerships in Local Economic Development. *Local Government Studies*, No. 11, pp. 19–33.

Moore, C., Richardson, J. J., Moon, J. (1989). *Local Partnership and the Unemployment Crisis in Britain*. Allen and Unwin, pp. 1–163.

Nelson, J. & Zadek, S. (2000). *Partnership Alchemy – New Social Partnerships in Europe*. Copenhagen: The Copenhagen Centre.

Nidasio, C. (2004). "Implementing CSR on a Large Scale: The Role of Government." Paper presented at the research track 7: "Policy Making and the Role of Government," III Colloquium of the European Academy of Business in Society: "The Challenge of Sustainable Growth: Integrating Societal Expectations in Business." Brussels.

Perrini, F., Pogutz, S., Tencati, A. (2004). "Corporate Social Responsibility in Italy: State of the Art." Paper presented at the Workshop on Corporate Social Responsibility, organized by *Journal of Management Studies* at the University of Illinois, Chicago.

Perrini, F. & Tencati, F. (2005). "La Corporate Social Responsibility in Europa: sostenibilità dei modelli di sviluppo." In S. Rosci (Ed.), *Il grande salto sociale*. Naples: Controcorrente.

Pongsiri, N. (2002). Regulation and Public–Private Partnerships. *The International Journal of Public Sector Management*, 15(6–7): 487.

Porter, M. (1990). *The Competitive Advantage of Nations*. New York: The Free Press.

Rosdahl, A. (2001). *The Policy to Promote Social Responsibility of Enterprises in Denmark*. Discussion Paper from Host Country Expert. Danish National Institute of Social Research. Copenhagen, 17–18 September.

Sapir, A. (2005). "Globalisation and the Reform of European Social Models." Background Document for the presentation at ECOFIN Informal Meeting in Manchester, 9 September 2005. Brussels: Bruegel.

Tencati, A. (2005). *Economic, Social, Environmtental and Sustainability Performance. Evaluation and Reporting at the National Level: Frameworks and Indicators*. G. Pivato, Department of Management, Bocconi University, and SDA Bocconi School of Management.

Tencati, A. (2004). "Public Policies to Encourage CSR: The Italian CSR–SC Project." Paper presented at the III Transatlantic Business Ethics Conference. Barcelona: ESADE.

Tencati, A., Perrini, F., Pogutz, S. (2004). New Tools to Foster Corporate Socially Responsible Behaviour. *Journal of Business Ethics*, 53(1–2): 173.

United Nations (2002). "The Social Responsibility of the Private Sector and Forcing Partnerships for Social Development." Report of the Expert Group. Meeting on National and International Cooperation for Social Development. Commission for Social Development.

United Nations Development Programme (2001). *Human Development Report 2001. Making New Technologies Work for Human Development*. New York: UNDP.

Wilson, A. (2002). Governments, Business and Corporate Social Responsibility: Towards a Partnership. *MT Eliza Business Review*, Winter–Spring.

Ysa, T., Lozano, J. M., Albareda, L. (2004). "Governments and Corporate Social Responsibility (CSR)." Paper presented at the III EABiS Colloquium, Ghent.

Zadek, S. (2001a). *The Civil Corporation: the New Economy of Corporate Citizenship*. London, Stirling, VA: Earthscan.

Zadek, S. (2001b). *Third Generation Corporate Citizenship*. London: The Foreign Policy Centre and AccountAbility.

Zadek, S., Hojensgard, N., Raynard, P. (2001). *Perspectives on the New Economy of Corporate Citizenship*. Copenhagen: The Copenhagen Centre.

Zadek, S. & Swift, T. (2002). *Corporate Social Responsibility and the Competitive Advantage of Nations*. AccountAbility; The Copenhagen Centre.

Zadek, S., Sabapathy, J., Dossing, H., Swift, T. (2002). *Corporate Social Responsibility Clustering–Leveraging Corporate Social Responsibility for National Competitive Advantage.* AccountAbility; The Copenhagen Centre.

Zadek, S., Sabapathy, J., Dossing, H., Swift, T. (2003). *Responsible Competitiveness. Corporate Responsibility Clusters in Action.* AccountAbility; The Copenhagen Centre.

Zappal, G. (2003). *Corporate Citizenship and the Role of Government: The Public Policy Case.* Australia, Research Paper, No. 4, 2003–2004.

Frank Dixon
Global System Change
New York

Sustainable Systems Implementation

Business has done great work, but its greatest work lies ahead. Responding to rising environmental and social problems, Wal-Mart, General Electric and many other firms have implemented aggressive corporate responsibility strategies. Done well, these strategies enhance profitability and competitive position, while improving environmental and social performance.

Yet in spite of this great work, environmental and social conditions are declining rapidly in many areas, indicating humanity is becoming more unsustainable, not less. To reverse this situation, ensure ongoing business prosperity and secure the well-being of future generations, a higher level of work is needed.

Without intending to do so, economic and political systems create conflicts between what's best for business and what's best for society. These conflicts drive environmental and social declines as well as create growing problems for business. In effect, modern systems make it impossible for firms to operate in a fully responsible and sustainable manner. Sustainability can only be achieved by improving these systems.

A sustainability approach called *Total Corporate Responsibility* (TCR®) combines traditional corporate responsibility efforts with system change efforts at the mid-level and high-level. Mid-level system change is focused on specific sectors, stakeholder groups, or environmental or social issues. High-level system change is focused on improving overarching economic, political and social systems. *Sustainable Systems Implementation* (SSI) is the high-level system change component of TCR.

Through SSI, committed leaders and system change experts work together with larger society to find practical, reasonable ways to evolve human systems into sustainable forms. Many good system change ideas and programs already have been developed. Most have low implementation rates. A key focus of SSI is taking these to much higher levels of implementation. The overall goal is to help build a more prosperous and sustainable economy and society.

TCR and SSI are being implemented by Gazeley Ltd, an independently operated Wal-Mart subsidiary based in the UK. Gazeley is a global real estate developer and sustainability pioneer that provides environmentally-superior distribution warehouses to Wal-Mart and many other clients around the world.

High-level system change is extremely complex, but also extremely important because sustainability is not possible without it. The largest problems facing business and society can be solved only by adopting a system perspective and using collaborative approaches. This article provides a summary of why high-level system change is needed and how business can use the collaborative SSI process to bring it about.

1. Sustainability Requires Systems Thinking

The primary purpose of capitalism and democracy is to enhance society. Clearly these systems do this in many ways. However, they also produce unintended consequences. Providing many people with high-quality, affordable products and services drives increasing environmental and social impacts.

As an expanding population with rising living standards consumes more resources and produces more pollution, environmental systems that keep humans alive are being rapidly degraded in many areas (ie: clean air, clean water, fertile soil, stable climate, forests, wetlands, aquifers, fisheries, biodiversity). In addition, growing social unrest and turmoil in many regions are being driven by a widening gap between rich and poor, disruption of traditional lifestyles in developing countries, and increasing emptiness and life dissatisfaction in developed countries (as indicated by rising anti-depressant drug use, obesity and compulsive behaviors).

The main reason our well-intentioned systems produce unintended consequences is the failure to think systemically. As quantum physics and common sense show, everything on the Earth is part of one interconnected and interdependent system. It is difficult for the human mind to contemplate all the details of the whole Earth system at once. As a result, human systems (economic, political, social) were developed in simplified, reductionistic ways that exclude many relevant, interconnected aspects of society, thus creating unintended consequences.

As it was when other human systems prevailed (feudalism, slavery, communism), we do not see the flaws of our thinking because we are too close to them. Many ideas that appear to be correct at the individual person, company or country level are actually counter-productive at the system level. (It is important to optimize first at the whole system level because individuals ultimately cannot prosper if the system that supports them does not prosper.)

In the economic area, there are several ideas that seem logical at the individual level, but are not logical at the system level. These economic system flaws include the failure to incorporate relevant environmental and social costs (externalities) into prices, which drives major price distortions, waste and inefficiency in our market system. The pursuit of ongoing economic growth in the finite Earth system makes it difficult for human society to achieve stability and balance. Time-value-of-money, a foundational economic concept, says that people and resources beyond 50 to 100 years are worthless. Therefore protecting them would be a foolish economic decision. And by focusing primarily on only one aspect of society's success (economic growth – GNP), other aspects are not adequately addressed. This causes the overall well-being of society to decline.

Political/regulatory system flaws include inappropriate influence of regulators and limited liability structures. The ability of regulated entities to influence regulators through campaign finance, lobbying and other processes makes it effectively impossible for government to hold firms fully responsible for negative impacts. Limited liability structures provide unlimited upside to investors, but cap the downside by transferring risk mostly to low and middle-income taxpayers. This is a major factor driving the widening gap between rich and poor.

Social system flaws include allowing the public to be misled by advertising and media. Powerful emotional advertising often misleads the public into feeling non-material needs for self-esteem, love and connection to others will be met through consumption. Also, media often is used in ways that mislead the public about issues that might hurt profitability, such as climate change.

These results are unintentional or secondary. The primary purpose is not to mislead the public or allow irresponsible behavior. It is to achieve success on the one prosperity metric we do effectively measure and manage – profitability, and by extension economic growth.

2. Going Beyond Conventional Corporate Responsibility

The above system flaws create conflicts between business and society, primarily by not holding firms fully responsible for negative environmental and social impacts. In a closed system, such as our Earth system, there is no free lunch. When firms negatively impact the environmental and social realms, feedback loops return these impacts and cause problems for firms, such as market rejection, lawsuits, activist campaigns, and opposition to opening new stores and facilities.

As the scale of economic activity increases in the finite Earth system, feedback loops shorten and firms receive pushback for negative impacts more quickly. This causes environmental and social issues to become more financially relevant to investors and firms. Growing financial relevance is driving the mainstreaming of corporate responsibility (CR) and socially-responsible investing.

Traditional CR strategies go beyond regulations and seek to reduce the negative environmental and social impacts for which firms are not held responsible. (Regulations often require firms to eliminate some of their pollution and other environmental and social impacts, but not all.) In thousands of cases, voluntary CR has improved profitability and competitive position by enhancing reputation, brand value and employee morale, reducing energy, materials and waste disposal costs, facilitating siting approval and access to new markets, and generally securing the right-to-operate.

However, voluntary CR rarely, if ever, allows firms to fully mitigate negative impacts and thus act in a fully responsible and sustainable manner. Beyond a certain point, companies attempting to fully mitigate impacts see costs increase relative to firms that are not fully mitigating. If a company tries to act "too responsibly" by going too far beyond regulations and attempting to mitigate all of its negative impacts on society, it will put itself out of business.

This is why it's important to hold firms fully responsible for negative impacts (hence the name Total Corporate Responsibility). In a competitive market, firms cannot act in a fully responsible manner and remain in business. Holding firms fully responsible in a practical way over time will make full responsibility and sustainability the profit-maximizing path.

Traditional CR approaches go beyond regulations and mitigate negative impacts until it becomes unprofitable to do so. Going further would violate the obligation to maximize shareholder returns. Through this mechanism, firms are compelled to degrade the environmental and social systems that support the economy and society.

If we cannot evolve economic and political systems into forms that better serve business, investors and society, their cumulative negative impacts will force system change, almost certainly in a disruptive manner. Given the rapid escalation of global environmental and social problems, it is highly likely that disruptive system change will occur sooner rather than later. If we wait until the issues are fully upon us, we will have few, if any, options. To effectively improve human systems, we must begin to address the issue in the near-term. While there may be no short-term solutions, the need to begin developing and implementing solutions has become a short-term issue.

The complexity of improving economic and political systems is daunting to everyone. But it is least so to business leaders. Business is in the best position to lead the system improvement effort because its experience operating in economic and regulatory systems gives it much practical knowledge (those who understand systems best are best able to improve them). In addition, with its vast innovative and creative potential, business has the ability to develop and implement successful system change strategies. Finally, business effectively is the most powerful human force on the Earth. Voluntary system improvement almost certainly will not happen if business does not drive it.

The key question is, how can business effectively lead the effort to improve human systems, while simultaneously performing well in existing systems? That is exactly what the SSI is designed to do.

3. Sustainable Systems Implementation

The goal of SSI is to make progress on the most complex challenge facing business and society – sustainability and the high-level system changes needed to achieve it. The challenge is so great that no single person knows the answer, though many have parts of the answer. Successful evolution of human systems into sustainable forms can only be achieved through a collaborative process, such as SSI. The most important requirement of SSI is to just begin, to make some progress. Once on the path of system improvement, clarity will increase and additional answers and actions will emerge.

SSI brings together leaders with the power to drive change from business and all other segments of society, along with experts that have good system change ideas. Many great system change ideas have been developed. Most have low implementation rates. SSI seeks to quickly drive good system improvement ideas to far higher levels of implementation. The SSI process involves planning and implementation. But since there already are many good ideas developed but not widely implemented, the focus of SSI is on implementation and action (hence the name Sustainable Systems Implementation).

The following shows SSI principles, structure, participants and management along with potential work focus, desired process outcomes and quick wins. Since SSI is a collaborative process, specific strategies and goals will be developed and agreed as the program progresses.

4. Principles

The following principles show important activities and viewpoints needed to evolve human systems into sustainable forms.

Systems Thinking. As noted above, the main driver of humanity's unsustainability is the failure to think systemically, which essentially means the failure to consider all relevant factors. Achieving sustainability by evolving human ideas and systems into sustainable forms can only be done through systems thinking. This greatly increases complexity because nearly everything is relevant from a systems perspective. Few things can be done effectively in isolation. For example, successful evolution of our economic system probably can be achieved only if done in tandem with evolving political, social and other systems.

This is difficult, but not impossible. Different systems thinking approaches could be used. For example, one could postulate a global mind and imagine how it would develop a sustainable economic system for the planet. Actually, this already has been done by nature. The model for a sustainable economy is all around us in nature. When seeking to address some of the most complex economic system flaws, such as externalities, limits to growth, time-value-of-money and measurement of success, one could ask, how does nature implicitly handle these issues.

One of the main reasons many good system change ideas have low implementation rates is the difficulty of understanding and addressing all relevant factors. By taking a systemic approach, all barriers to success can be effectively addressed. These barriers often will include lack of public awareness and regulations that make responsible corporate behavior impossible.

Simplicity. The complexity of system change is so vast that great efforts must be made to achieve clear, simple communication. One of the greatest barriers to system change is the lack of public awareness about the systemic drivers of environmental and social problems along with the lack of public will to address systemic issues. Rendering discussions down to simple terms and concepts will greatly facilitate the expansion of public awareness. For example, everyone believes our children and generations beyond them should survive and prosper. This is a simple, unifying concept. Clearly showing how current economic and political systems threaten our childrens' future can motivate action.

Inevitability. Often during discussions about how to change systems, concerns about difficulty and complexity are used intentionally or unintentionally as excuses for inaction. This argument for inaction usually results from the perspective that systems could remain the same. From this perspective, system improvement work might seem unnecessary.

However, as we better understand the dynamic, ever-changing nature of human society, especially when it is in conflict with natural systems, we realize that not changing is not an option. There are only two options – voluntary or involuntary system change. As we then understand that involuntary system change will be disruptive to business and society as well as unfair to future generations, common sense dictates we really have only one option. Understanding the inevitability of system change greatly increases the willingness to deal with complexity and the commitment to take action.

Evolution not Revolution. SSI is intended to evolve human systems before the negative impacts of current systems force traumatic, possibly revolutionary change. The evolutionary approach to system change is based on the idea that committed leaders working together with larger society can find practical, reasonable ways to evolve our systems into sustainable forms. The goal is to do what humans always do – improve – not to go backwards to systems that didn't work, but to combine ideas from the past that did work with new ideas, then develop something new and better. This approach will maximize the well-being of business and society, while minimizing disruptions as the transition to sustainable systems occurs.

Be Willing to Question Everything. To achieve meaningful system change, we must be willing to question all human ideas and systems. In the past, many systems intended to benefit society turned out to be not beneficial, even destructive. It would be illogical (and possibly arrogant) to assume we are not making similar mistakes today. Human ideas and systems are always evolving. No idea or system (economic, political, social) should be seen as unquestionable, as having any inherent right to exist. One could say that the only things with an inherent right to exist are life, that which supports life (the environment), families, communities and the rights of people to prosper and be secure. Human systems should exist only if they protect and serve life over the long-term. If they do not, they should be evolved into forms that do.

Abide by Nature's Laws. Nature is the most powerful force on the Earth. Humans have no option to live outside the laws of nature. One could say there are laws of nature on the inside (basic human rights) and on the outside (environmental restrictions). To the extent we operate outside these laws, we will be adjusted until we do comply. That is why we must be willing to question everything. This is not a philosophical statement. It is simple common sense and logic. In reality, it does not matter what's happening in the human mind. Nature will do what nature will do. Reality will always prevail. If our ideas and systems are not aligned with nature's laws, they will disappear, as have all other human systems that weren't aligned with nature and reality.

The beliefs of many indigenous groups caused their actions to be aligned with nature. They often believed that nature, as the sustainer of all life, should be revered and respected, rather than seen simply as a source of raw materials. Given the effectiveness and sustainability of many indigenous ideas and economic practices, these should be considered when seeking to improve modern human systems.

Non-Judgment. It is easy to become angry when one considers the destruction of environmental life support systems, inequity and unhappiness in the world, especially when one considers what this means for our children. But anger often is not the appropriate response, in part because it can polarize different groups and block progress. In a sense, we are like children learning to walk. Our collective actions show that we (society in total) have not learned how to live sustainably on this planet yet.

The advancement of our ideas and systems has not kept pace with the rapid advancement of technology. For example, we produce many substances never seen before in nature (nano-materials, genetically-engineered foods, more than 80,000 synthetic chemicals). Many of these tend to disperse into the land, air and water, then accumulate in humans, causing growing reproductive problems, cancer and other illnesses. The large majority of these substances has never been tested for safety by an independent third party, in part because the firms making them are able to influence the regulatory process. In addition, we are removing millions of years of fossil fuels and other resources from the Earth's crust, leaving little for future generations but pollution.

Business and political leaders are criticized often for these and many other environmentally and socially damaging activities. However, none of these leaders intends to damage society. They are well-intentioned people operating in systems that often compel them to do the wrong thing for society. The enemy is not individual firms or leaders. It is our short-sighted ideas and resulting systems. In the same way that we would not criticize a child for not knowing how to walk, we should not criticize business for doing what systems compel it to do. Instead, we should work together to improve our ideas and systems.

Don't Expect Individuals to Change. The purpose of SSI is system improvement, not unilateral change. There should be no expectation that individual persons or firms will unilaterally modify their activities. (There are many good programs promoting individual change, including conventional CR. But that is not the purpose of SSI.) Most people are focused on just surviving. They do not have the luxury of considering longer-term issues like sustainability. Some people will adopt environmentally and socially responsible lifestyles, even if it costs more, because they feel it's the right thing to do. Most will not.

As systems become sustainable, the full cost of environmentally- and socially-damaging products and business practices will be included in prices. This will cause the responsible lifestyle to become the easiest and least expensive lifestyle. Also, as noted above, firms often cannot act more responsibly unilaterally and remain in business. SSI is intended to drive system improvements that encourage all people and firms to act more responsibly.

Carrot not Stick. SSI is not intended to take anything away from anyone or force individuals and firms to change. Instead, the goal is to develop more appealing and effective systems that draw citizens and firms toward more responsible activities. Better systems will allow more satisfying lifestyles while producing more sustainable business performance. This approach is far more complex because all voices must be heard and considered. In practice, some firms not committed to full responsibility might resist change. But the principle of developing appealing systems with reasonable transition strategies should minimize resistance to system change.

Responsibility. Probably the most important flaw of modern economic and political systems is the failure to hold firms fully responsible for negative environmental and social impacts. When this occurs in a competitive market, firms cannot fully mitigate impacts and remain in business. Holding firms fully responsible aligns what's best for business with what's best for society. It gives firms the incentive to act in a fully responsible and sustainable manner by making such action the profit-maximizing path.

More responsibility does not mean more regulations. Firms often complain about regulatory burden. Much of this burden is due to inefficient, cumbersome regulations resulting from inappropriate influence of the regulatory process. As this influence is removed and as the higher principle of full responsibility prevails, simplified and far more effective regulatory strategies can be implemented, especially if firms are given the flexibility to figure out how to achieve full responsibility. As noted above, not holding firms fully responsible compels negative impacts, which creates growing problems for investors and firms. Holding firms fully responsible in a practical way over time will benefit business and society.

Focus on Results. While dialogue, planning and developing systemic solutions are key aspects of SSI, the strong focus is on generating quick wins that build enthusiasm and support for addressing longer-term, more complex issues.

Simultaneous Success. SSI is intended to enhance current business performance in existing systems while proactively positioning firms for success as more sustainable systems are implemented. To minimize costs, SSI participation requires relatively little management time. All actions developed under SSI should be practical and support simultaneous success. Some actions can be designed to take advantage of participants' strengths, for example by promoting regulatory changes that build competitive advantage.

Visionary, Courageous Leadership. Business leaders operate in fixed systems and structures. Contemplating how these systems might be improved then working to improve them involves a high degree of uncertainty. It is far easier to remain safely within existing systems. It takes courage to step back, look at the big picture and evaluate how overarching systems create problems for business and society. Business can benefit society greatly by leading the effort to practically and reasonably improve human systems. But it will only happen if business leaders have the courage and vision to make it happen.

5. Structure and Participants

Initial SSI efforts will focus on the national level since overarching economic and political systems largely are created and maintained at this level. However, as groups of stakeholders within a given country consider how they might evolve their systems into sustainable forms, they'll be confronted with the reality that they must operate and compete in larger global systems. As a result, once several national and multi-national SSI efforts are underway, these groups would spend part of their time working together on a global SSI effort that seeks to improve the highest-level economic and political systems.

The SSI approach could be used at any level. However, it is most important to work at the national and international levels since the purpose of SSI is driving high-level system change. In addition, focusing at this level is important because rapid, creative national and international responses are needed to effectively address escalating environmental and social issues.

In no way does SSI intend to minimize the importance of individual, community and regional sustainability efforts. To become sustainable, human society must emulate and live within the laws of nature. Nature mostly operates in local, self-sustaining communities. The building blocks of a sustainable world almost certainly will be sustainable communities. There are many excellent sustainable community and regional efforts underway around the world. These would be important participants in national SSI efforts.

There also are many excellent economic and political reform efforts focused at national levels. While these are having some impact, successfully evolving our systems into sustainable forms requires a much greater impact. By combining powerful leaders with system change experts, often from these groups, SSI is intended to achieve this impact.

To ensure that all voices are heard and all good ideas considered, leaders and experts from all segments of society will be invited to participate. SSI will be open to segments including business, government, military, academia, non-governmental organizations (NGO) and media along with civil society groups representing the elderly, women, youth, minorities, religions, communities, labor and any other group interested in constructively and collaboratively participating.

6. Convening and Managing

Convening the SSI involves identifying initial participants. Ideally, the initial group would include highest-level leaders from business, government and civil society. Based on initial feedback, many NGO's and academic institutions would be interested in participating in and possibly co-convening the SSI. For NGO's with an environmental or social mission, systemic issues probably are the primary driver of the issues they seek to address. For academic institutions, evolving human ideas and systems into sustainable forms through interdisciplinary, systemic, collaborative approaches represents the most interesting and important academic challenge.

Convening and managing national SSI efforts will require support, including recording and analyzing proceedings, keeping participants informed and in communication with each other, providing venues, planning and managing SSI events and forums, initiating and managing implementation of quick wins and other projects, and providing funding for these and other activities. Given strong mission alignment, it is likely that some NGO's and academic institutions will provide these types of support.

Potential participants will be contacted to discuss the importance of the SSI, benefits of participation, general process, and potential work focus and outcomes. In the early stages of SSI, it will be very important to include people who are deeply committed to sustainability – to improving human systems in ways that are responsible and highly beneficial to society.

An interdisciplinary approach will be required to improve human systems. Human knowledge within disciplines is deep, but across disciplines it is much less so. Unwillingness or inability to address issues outside ones area of expertise is a primary reason good system change ideas often do not get implemented widely. As a result, it is critical that early participants be open-minded and willing to consider others' opinions and new approaches. People with inflexible agendas would limit success. As SSI gains momentum and resilience, all views and ideas can be considered and integrated to the extent practical.

The first meeting should be kept small (about ten to fifteen people) and held in a confidential, closed setting. This will facilitate open dialogue and consideration of complex issues. It might be difficult for leaders to address some issues in a public setting, for example, balancing the demand for ongoing economic growth with what's best for society.

To facilitate dialogue during the first meeting, possible quick wins and systemic issues to be addressed will be suggested. One or more expert presentations will focus on high-level systems thinking that integrates all parts of society and nature. As the SSI progresses and specific work areas emerge, experts in these areas will be asked to propose and help implement specific solutions.

Examples of potential desired outcomes of the initial SSI meeting might include the following:

- Ensure awareness of the need for system change among participants and secure strong commitment to achieving it.
- Reach general agreement on the major systemic issues that must be addressed.

- Begin to develop a vision of a sustainable world that helps guide system change efforts.
- Generally outline short-term and long-term system change actions needed.
- Commit that the group will continue to work together to achieve system improvements.
- Agree on other parties needed for success.
- Identify quick wins and ways to achieve them.
- Commit to achieving quick wins.

7. Work Areas

The SSI is intended to be a long-term collaborative effort involving many different high-level system change activities organized by a unifying vision. The founding participants will begin to chart the course and build momentum for this most important work. The work of the SSI could be divided into three areas: Framing, Quick Wins and Longer-Term Challenges and Opportunities.

7.1 Framing

SSI framing work involves understanding the context and developing system improvement plans. Framing activities include developing a vision of a sustainable country or world, honestly identifying the current state of the world (i.e.: being clear about environmental and social trends and likely impacts on business and future generations), identifying systemic barriers to sustainability (i.e. what are the major ideas, system flaws and other factors blocking movement from the current to the desired state), and developing plans to overcome barriers and achieve sustainability. Traditional strategic planning approaches applied at a larger scale, such as SWOT (Strengths, Weaknesses, Opportunities, Threats), would facilitate this work.

Developing a vision of a sustainable world (or country, depending on the level of focus) may be a small part of the overall SSI effort. But it is a critical component. All strategies need a clear focus. Without one, the strategy usually fails. A major issue for the sustainability movement is the lack of a clear vision. Sustainability is considered by many to be a nebulous term with several different definitions. Developing a clear vision will help in several ways.

Having a clear vision of where we'd like to go clarifies major barriers to success and helps to prioritize necessary actions. Focusing on the end point, for example prosperity for future generations, facilitates finding common ground, overcoming differences, building consensus and dealing with complexity. While working on system improvement, one could easily get lost in overwhelming complexity and details. The vision serves as a constant reference point throughout the process. One often can find clarity when mired in complexity by stepping back and asking, what's the overall purpose here.

The vision also facilitates systems thinking by helping people see the big picture of human society over time. From this perspective, it's clear that things are not as complex as we might have imagined. We see that human ideas and systems are always evolving. From the human perspective, things can seem to be stuck. But they never are. They are always in a process of changing, regardless of how it might seem to the individual. This helps us to let go of the idea that things will stay the same. It allows us to think more creatively about what will come next and how we might shape that change.

Having a clear vision of a sustainable world that many, if not nearly all, agree on will build strong commitment to sustained action. While a sustainable world could take an infinite number of forms, there are many factors that most would agree on. For example, there should be clean air, land and water, now and ten thousand years from now. Basic rights should be protected and basic needs met. Most importantly, our children and those after them should prosper on every level.

Finally, having a clear vision of where we'd like to be helps to clarify the limitations and inaccuracies of our current ideas. Human systems are the fruit of ideas. We probably won't be able to evolve our systems into sustainable forms unless we are willing to question our beliefs and find the courage to adopt new ideas. The ultimate driver of humanity's unsustainability is our individual and collective beliefs and worldviews. Recognizing this, we can begin the large and possibly long process of developing more realistic and sustainable worldviews, ones that address the appropriate rights and role of humanity on this planet.

7.2 Quick Wins

The SSI involves several levels of action occurring simultaneously. Since the approach has a strong bias for action, the SSI group would produce an initial sustainable country or world vision, assessment of systemic issues to be addressed, and suggestions for short-term action. As these quick wins are pursued, vision refinement and analysis of more complex, longer-term issues, such as the system flaws mentioned above, would continue.

There are few quick wins in the system change area because most of the changes needed are very complex. Probably the most important requirement for achieving system change is moving from the discussion to the action phase. Experts have been discussing and developing good system improvement ideas for many years. The work required now is practical implementation. Just beginning to take action that has the potential to achieve broad system improvement could be seen as a quick win.

Along these lines, examples of Quick Wins that might be developed by an SSI group include:

- Identify one or more products used in large volumes by all or most of the group, then agree to buy them only in sustainable forms.
- Agree to provide sustainability awareness and training to employees.
- Support and publicize broader, more accurate measures of social well being than GNP.
- Support a coordinated media campaign focused on raising public awareness about the need for and benefits of system change.
- Agree to instruct lobbyists to request/support practical regulatory changes that hold firms more responsible over time, thus making further impact mitigation profitable.

- Support reforms that publicly fund political campaigns and eliminate the ability of any person or group to financially influence the political process.
- Seek practical tax code and regulatory changes that internalize the costs of environmentally- and socially-damaging products, thus making responsible products more competitive.
- Engage the capital markets in driving system change by supporting the launch of TCR investment funds that provide superior returns by shifting investments toward well-managed system change leaders.

7.3 Longer-Term Challenges and Opportunities

To achieve sustainability, many complex issues must be addressed, including population growth, inequity and the economic, political and social system flaws mentioned above. Addressing problems of this scale typically has not been the responsibility of business. However, these challenges are a growing concern for business because they are a growing concern for society. As society declines, business declines. The opportunity of high-level system change is to ensure ongoing business prosperity by ensuring the ongoing well-being of society.

Business is in the best position to lead the effort to address complex issues. Perhaps a main reason these issues remain unresolved is that business has not been adequately involved. The SSI is intended to create a vessel out of which effective solutions to these complex challenges can emerge. It is based on the idea that no person or group has the answer. But society, led by business, can develop ways of living on this planet that improve the economy while protecting future generations.

This section discusses some of the sources of complexity that must be addressed to achieve sustainability. Using the SSI Principles shown above will be critical to successfully addressing complexity and evolving systems into sustainable forms. To illustrate how SSI efforts might address complex, longer-term issues, a few hypothetical examples are provided. Specific issues and how they might be addressed will be decided by the SSI group.

Sustainability (and the high-level system change needed to achieve it) is the most complex challenge facing humanity. Nearly every other challenge is a sub-element of the sustainability challenge. Sustainability involves achieving long-term prosperity for humanity by living in balance with nature. Anything that significantly interferes with this is ultimately relevant. There are economic, political and social aspects to human society. These are all interconnected. Significant progress in any area probably can be made only by considering related, relevant factors in all areas.

In the same way that an individual has a mind and body that must be considered when seeking a successful life, human society in total has a collective mind and body. The collective human body has physical impacts on the world that drive environmental declines. The collective mind effectively is the collection of human beliefs and worldviews. These beliefs drive our individual and collective behavior. If we wish to halt environmentally destructive behavior, we must change our beliefs and worldviews in many ways.

As SSI groups consider sustainability challenges, the need for systems thinking will become apparent. Complex issues that can be resolved only through systems thinking include externalities, limits to growth and population growth. Regarding externalities, the price consumers pay for gasoline does not include the full costs of delivering and using it. Costs not included in gasoline prices include increased public and private healthcare costs (due to tailpipe and other emissions), military costs of ensuring oil supply, costs to maintain highways and bridges, lost work time related to traffic jams and lack of public transportation (resulting from subsidized fossil fuel prices), storm damage related to global climate change, reduced quality of life, and many other factors.

By charging an artificially low price, fossil fuels are over-consumed, which drives numerous environmental and social problems. Charging the correct price for fossil fuels probably is the single most important short-term action we could take to largely reduce global climate change and many other environmental problems.

But it is not that easy. There are strong political and social barriers to charging the correct price for fossil fuels. If an SSI group chose to focus on subsidized fossil fuel prices, it would seek to address these systemic factors. For example, the group might help consumers understand that the real cost of gasoline is much more than what they pay at the pump. Costs also includes higher income taxes and medical costs and lower quality of life.

Incorporating full costs into prices would substantially lower the real cost of gasoline because consumers effectively would be paying to prevent illness and other problems, rather than paying to clean them up, which is nearly always far more expensive. If this issue were made clear to consumers, they would understand that a 100 percent increase in gasoline prices might, for example, mean a 50 percent reduction in total out-of-pocket costs, when income tax, healthcare and other factors are included. Charging the correct price for fossil fuels would shift taxes from income to pollution. The tax code would be used to motivate positive behavior (reducing pollution) rather than negative (discouraging job creation).

But even making consumers aware of the real cost of fossil fuels probably would not be enough to allow charging the correct price. It most likely would require considering other political and social issues. These might include questioning the wisdom of organizing society in a way that requires so much travel, the decline of village and community life driven partly by a movement toward a global mono-culture based on consumption, the obligations of society to meet the basic needs of all, and the short-term focus of government.

Another example of high complexity involves improving an economic system that encourages infinite growth in a finite system. A more sophisticated system would recognize that material growth must stop at some point, as it always does in nature. As limits in the finite Earth system are approached, the focus would shift from growth to maintaining balance with nature, while more effectively measuring, managing and fulfilling non-material and material human needs.

One of the most difficult systemic challenges involves population growth, a major driver of unsustainability. Using the SSI Principles will help in addressing this and other complex issues. For example, using the principle of Inevitability, one sees that human population will be limited by nature, as it always has been for any other creature that overshot the environment's ability to support it. This principle could be interpreted to mean, once it becomes clear that population growth (or any other factor) threatens the well being of our children, then not acting is not an option. In other words, we must take action regardless of the difficulty or complexity involved.

Another principle, Carrot Not Stick, would be helpful in addressing population growth. Some might suggest that family size should be limited. But this would raise many valid concerns. Pressure to reduce family size might work in the short-term, but create other problems. The more difficult, but likely to succeed approach is to consider the wide range of relevant factors through a collaborative approach. Relevant factors might include ownership and management of local resources, the amount of life a given piece of land could support, introduction of external values and belief systems that disrupt previously sustainable lifestyles, strengthening communities in ways that better meet human needs as an alternative to larger families, and internalizing costs so that the real costs and inefficiencies of the global food production system become apparent.

Clearly the complexity of systemic issues is immense. Rather than solving these problems, the goal of SSI is much less ambitious. It seeks to convene groups of leaders and experts who will apply practical systems thinking to these issues. As said above, the goal simply is to just begin, to make some progress on system improvement. Done well, the SSI will catalyze broader system improvement efforts in larger society.

Language will be key to achieving success. Business leaders probably should not question some system change issues directly. For example, the stock market would not react favorably to a CEO questioning the wisdom of ongoing growth in revenues and earnings. That is why this type of statement would never be part of an SSI approach. Instead, a group of leaders might collectively question the wisdom of an economic system that puts business in conflict with society. It would clearly show how these conflicts hurt business and investors. Then it would suggest that many parties must work together to build a better system, one that protects business and investors by protecting society.

8. Benefits to Business and Society

Business leaders usually are consumed with meeting investor, market and other demands. Taking time to step back and look at the big picture can be highly beneficial. As one adopts a higher, whole-system perspective, it becomes clear that much of what was true in the past will not be true in the future.

We live in times of unprecedented, rapid and accelerating change. One factor that's changing rapidly is the pace at which firms get pushback for their negative environmental and social impacts. As this continues to accelerate, corporate responsibility will become the primary driver of business success in the 21st Century.

Conventional corporate responsibility approaches do not allow firms to fully mitigate impacts. As a result, pushback from society is inevitable. SSI enables firms to collaboratively and practically address systemic issues that prevent full mitigation. Like more traditional management approaches, such as Zero Defects and Total Quality Management, SSI adds value by driving creative, practical, often incremental improvements. Working with others, firms lower impacts in ways that enhance profitability and competitive position.

Due to high complexity, no firm has effectively addressed high-level system change. Yet system change is the most important sustainability issue because sustainability is not possible without it. This presents a major opportunity. Companies aggressively working in this area will be seen as the most visionary and courageous – the true sustainability leaders.

By initiating the SSI in the UK, Gazeley is advancing its already strong reputation as a sustainability leader. The firm will benefit from SSI activities that raise customer and public awareness about the financial and environmental benefits of Gazeley structures and services. Through collaborative work with government, NGO's and others, regulatory reforms will be sought that favor Gazeley's environmentally-superior distribution warehouses, thus building competitive advantage. Gazeley personnel will benefit from the knowledge that their firm is pioneering the most advanced strategy for protecting their children.

Pressure from the financial community to supply ever-increasing financial returns obscures the fact that the primary obligation of firms is to society, not to shareholders. Companies do not have an inherent right to exist. That right must be earned. Providing valuable products and services cannot be used as a justification for ongoing negative impacts, as it often is. Even if regulations allow firms to negatively impact society, as they often do, firms have an obligation to go beyond this and mitigate all impacts. SSI provides a practical means of fully mitigating impacts over time.

Through SSI, business leaders can take a seat at the system change table. Rather than having system changes for which they are not prepared wash over them, leaders can help to shape practical, reasonable changes that improve business and society. Even on a purely financial basis, this makes sense. As environmental and social issues become more financially relevant, the fiduciary obligation to maximize shareholder returns demands that firms mitigate negative impacts to the greatest extent possible.

We all want our children and grandchildren to be proud of us. We all want to provide a prosperous, healthy, sustainable world to future generations. Let's do whatever it takes to make this happen!

Laszlo Fekete
Business Ethics Center
Corvinus University of Budapest

Public versus Private Domain: Knowledge and Information in the Global Communications Network

It goes without saying that knowledge and information are the most valuable commodities in the new economy. Though knowledge and information as private goods could provide great business opportunities for rights holders in the global communications network, they exhibit the distinctive characteristics of public goods (Samuelson 1954, pp. 387–389; Stiglitz 1999, pp. 308–325). Therefore, the commodification of knowledge and information requires a strict proprietary regime which restrains free access to them and enforces effective legal protection over their production, use, and dissemination. If the accessing and using rights of the individual users were free and unlimited the legal entitlements of rights holders would be worthless.

The pervasiveness of the new information and communications technologies as powerful learning and knowledge sharing systems and the digitalization of knowledge and information goods facilitate their production, use, and dissemination, and at the same time make difficult and expensive to enforce effective proprietary regulation and control over them. In addition to the difficulty and the high transaction costs of the legal enforcement of proprietary rights, the social climate does not seem to be particularly supportive to the propertization of knowledge and information in the global communications network. In spite of the current trends of restrictive legislation and jurisdiction as well as of expansive and unprecedented private legislation of the rights holders, society persistently tends to believe that knowledge and information mainly belong to public goods and resists accepting their growing private appropriation and effective proprietary control over their production, use, and dissemination. Briefly, knowledge and information are usually conceived as a common pool of symbolic resources for the cultural reproduction of society. So, people are not willing to pay for knowledge and information goods to regain what they believe to be rightfully entitled to know, use, and contribute to. Without the empowerment of the possession and exercise of these individual rights and freedoms, people merely are kept aloof from becoming the autonomous members of political-cultural community.

The rights holders also endeavor to control the flows of all forms of computer-mediated contents by means of the private ordering of accessing and using rights of users in combination with copy-protection technologies, digital rights management systems, platform dependent applications, micropayment system, zoning, and so forth (Benkler 2006, pp. 397–459). They go well beyond the initial rights and legal entitlements originally assigned to them by the law and habitually infringe the basic constitutional rights of the users, like freedom of expression, the right to privacy, and the right to fair trial by taking advantage of the opportunity of forum shopping (Balkin 2004, pp. 19–22; Boyle 2000, pp. 345–350; Netanel 2000, 1879–1886; Walker 2003, p. 24). These private encroachments on the users' individual rights and liberties further incite discontent and resistance. It is not surprising that hackers, cyberpunks, outlaws, and code breakers are usually regarded as public heroes and heroines in urban folklore who fight the enclosure of the public domain and the infringement of constitutional rights and liberties of the users. They are rarely stigmatized as villains therein.

1. The Tragedy of the Commons or the Anticommons?

Besides this common belief, many legal scholars, philosophers, scientists, and social scientists also emphasize that knowledge and information are social and cultural products made, shared, settled, and revised in democratic discourses, open scientific debates, and the pragmatic self-understanding of society. Therefore, the basic notions of mainstream economic paradigm about scarcity, exhaustibility, rivalry, and excludability, which are the distinctive characteristic of tangible goods, can be hardly applicable to the production, use, and distribution of knowledge and information (Kaul et al. 1999). In some respects, knowledge and information are not fit into the framework of neoclassical economics. Each individual can maximize the use of knowledge and information goods without exhausting the original resources, passing an excessive cost burden onto others, leaving anybody worse off than before, or excluding anybody from parallel exploitation and enjoyment. The overexploitation of knowledge and information does not bring about economic shortage and social threat; meanwhile, their underexploitation could lead to economic backwardness and social degradation (Vanneste et al. 2004, pp. 13–14). An open-access regime does not have inevitably harmful effect on social welfare, artistic, cultural, and scientific advancements as some law and economics scholars endeavor to argue against the public domain referring to Demsetz's theory on the impact of externalities in the development of property right system and Hardin's popular metaphor about the tragedy of the commons (Demsetz 1967, pp. 347–359; Epstein 1989, pp. 1488–1489; Hardin 1968, pp. 1243–1248; Landes and Posner 2003, pp. 471, 487–488). As a consequence of positive externalities and network effects, knowledge and information will never be exhausted under an open-access regime. In lack of rivalry, an open-access regime does not cause congestion or overcrowding in the use of knowledge and information, either. The opposite is the case: hedonistic and flamboyant behaviors in the consumption of knowledge and information goods are quite desirable. Therefore, converting Adam Smith's frequently quoted proposition, every prodigal man appears to be a public benefactor, and every frugal a public enemy in the production, use, and dissemination of knowledge and information goods. Under an open-access regime, knowledge and information will be continuously proliferated. Meanwhile, under a proprietary regime, the strict private control of the production, use, and dissemination of knowledge and the flows of information can cause scarcity, underprovision, inefficient resource allocation, endowment effect, holdup problems as well as deadweight social losses and cultural entropy (Heald 2007, pp. 35–41; Gordon 1992, pp. 153–163, 177–180; Netanel 1996, pp. 306–336; Posner 1992, pp. 277–278; Schultz 2002).

According to the mainstream economic paradigm, non-rivalry and non-excludability of knowledge and information goods are especially serious impediment which could frustrate the rights holders to recover production costs and to earn return on investments even if demands are sufficient and society attributes high cultural and economic values to innovative knowledge and novel information goods. For the reason that each additional user can consume knowledge and information goods, whether it is on-line newspaper article, scientific paper, symphony, or software once have been produced, at zero marginal costs, market itself is not a proper mechanism to set price above them. Knowledge and information as public, non-rival, and non-exhaustible goods are truly idiosyncratic to the established system of the market economy and the basic tenets of neoclassical economics. Therefore, knowledge and information as proprietary goods entirely rely on the existence of intellectual property laws and the effective legal enforcement of rights holders' proprietary claims. Indeed, the law itself transforms knowledge and information into commodities. By marking out the boundary corners of knowledge and information goods in the elusive fields of culture and staking out the legitimate claims of rights holders in terms of the scope and length of protection, governance, excludability, and exclusivity, the law makes knowledge and information scarce, rival, exhaustible, and excludable economic resources in order to recover the production and development costs and to ensure the economic gains and further commercial opportunities of private beneficiaries. The imposition of legal, judicial, and technological constraints on the production, use, and dissemination of knowledge and information goods serves the aims of the refutation of the basic feature of culture, science, and communication as collaborative enterprise and the reinforcement of the well-established division between producers and consumers (Barthes 1974, pp. 4–5). If the law provides individuals and business entities with proprietary rights and legal entitlements over the production, use, and dissemination of knowledge and information goods, non-owners' rights and freedoms will be inevitably circumscribed.

The rise of the global communications network as a new public forum for collaborative enterprises, creative endeavors, and information exchange is juxtaposed with the private appropriation of knowledge and information goods from the outset. The enormous success of the global communications network demonstrates that it can very efficiently fulfill the functions of production, use, and dissemination of knowledge and information. Since the global communications network has become a cornucopia of knowledge and information in the last fifteen years, it proves that the digital amplification and global accessibility of the public domain do not fade the spirit of innovation away. The exponential increase of cultural and technological innovation renders the well-established economic argument inapplicable, that is to say, exclusive proprietary rights over knowledge and information goods are necessary to create suitable incentives for owners to produce them and efficiently exploit their inherent economic values. Economic data do not support the fear of underproduction of knowledge and information goods owing to the liberal or relaxed intellectual property rights regime as the advocates of strict and extended regime complain. And what is more, its smooth and evolving operation is also feasible from the economic point of view. However, it still needs to be proven that the expanding propertization of knowledge and information goods – ranging from gene sequences and mathematical theorems to scientific databases, software algorithms, and cartoon figures – and the complete internalization of the benefits of their inventions and uses will enhance social welfare in general; their private appropriation and the creation and enlargement of exclusionary anticommons will further stimulate the amplification of their production, use, and dissemination (Balkin 2004, pp. 26–31; O'Rourke 2000, pp. 1178–79).

2. Public Goods, Private Goods, and Regulatory Givings

Paradoxically, the more our written, visual, and acoustic culture becomes public creative endeavors in the global communications network, the more efforts are made for withering away the public domain and fencing in knowledge and information goods. The attempts of turning knowledge and information into proprietary goods have been revealed in the profound changes of intellectual property law regime in the last two decades. The old intellectual property law regime was intended to strike the fair balance between the economic interests of the authors and the benefits of society. The early framers laid down in the British Statute of Anne (1710) and the Constitution of the United States (1788) that the primary aim of the copyright law was "the encouragement of learning" or "to promote the progress of science and useful arts, by securing for limited times to authors and inventors the exclusive right to their respective writings and discoveries". Especially, the latter document emphasized the primacy of social interests over authors' economic rights. The Berne and the Paris Copyright Conventions (1910, 1971) and the formerly enacted intellectual property laws had been gradually detached from these original intentions. Although they ceremonially recited the advancement of science, culture, learning, or the importance of public benefits in their preambles, they focused for the most part on the economic interests of rights holders. However, these copyright provisions in regard to the limited scope and length of protection, governance, and exclusivity were fairly generous with non-owners and future authors as compared with the current trends of legislation. The fair use doctrine, for instance, served the very purpose of intellectual property laws, namely, the encouragement of public engagement in cultural reproduction, learning, and creative works. In general, the provisions of intellectual property laws still reflected some concern of distributive justice among past, present, and subsequent generations of authors and of society.

Since the mid-1990's, the consecutive amendments enacted by the World Trade Organization, the World Intellectual Property Organization, and different national legislatures have significantly extended the scope and length of protection, governance, excludability and exclusivity and introduced more and more trade-centered and industry-specific – sometimes ephemeral and bizarre – provisions into intellectual property laws, which mandate the use of certain technologies and devices, inhibit, or entirely prohibit others. Today, three basic documents – the WTO Agreement on Trade-Related Intellectual Property Rights (1994), the WIPO Copyright Treaties (1996), and the Digital Millennium Copyright Act of the United States (2000) – establish the legal framework of intellectual property laws for national lawmaking processes worldwide. In the European Union, the Directive on Copyright in the Information Society has become the part of the acquis communautaire since 2001; though, a few member states, like Denmark and Hungary, alleviate some controversial provisions on technological protection measures in national law. In any case, the EU Directive on Copyright in the Information Society also follows the current trends of intellectual property legislation: intellectual property is mainly considered as trade-related matter for legislature, jurisdiction, and enforcement and framed in terms of economic law (Guibault et al. 2007, pp. 1–16).

The new provisions of these international treaties and laws strive for setting the course of future technological development as well as for determining the human use of knowledge and information goods in the global communications network biased in favor of the commercial interests of the present corporate rights holders. Contrary to the vindication of the early framers for striking a fair balance between the rights of authors and the interests of society, the persistent control of access to knowledge and information goods in the global communications network is the main focus of the prevailing intellectual property laws, today. International bodies and national legislatures are much more concerned with the effects of demand diversion and rent dissipation of corporate rights holders that may result from the global use of the new information and communications technologies and the progressive migration of knowledge and information to free zones of digital environment than the overall social and economic consequences of economic rent-seeking, endowment effect, holdup problems, and deadweight social losses for global welfare. The recent framers simply regard the free flow of knowledge and information in the global communications network as knowledge spillovers or externalities which generate economic values for third-parties and prevent rights holders to capture the full economic benefits of their rights and legal entitlements. They identify the intrinsically cumulative nature of research, innovation, information, and knowledge production with the problems of knowledge spillovers or externalities which should be overcome by means of exclusive property right arrangements, technological protection measures, and public legal enforcement. In the context of mainstream economic paradigm, the fair use provisions of the former intellectual property rights regime are explained as the unintended outcomes of market failures, namely, the high transaction costs of protection, enforcement, and bargaining over the terms and conditions of use prevent owners to internalize the full economic benefits of their rights and legal entitlements. Therefore, legislatures significantly circumscribed the breadth and scope of fair use provisions for sharing, recording, displaying, and storing digital contents in the whole corpus of the prevailing intellectual property laws and grant comprehensive control and authority over all manifestations of possible uses of knowledge and information goods to private and corporate rights holders (Benkler 1999, pp. 26–30; Birnhack 2006, pp. 505–517; Bradford 2005, pp. 7–15; Nimmer 2000, pp. 673–674; Nunzinato 2002, pp. 64–77).

The provisions of intellectual property laws in force and the new private ordering schemes throw off the well-established balance between the economic interests of rights holders and the benefits of society. In addition, the exclusive and almost persistent ownership rights of the present right holders impose significant limitations on present and future creators who will be impeded to build freely upon their predecessors' and contemporaries' ideas and progresses, and to incorporate some elements into their own works. As a result of abandoning the principle of intergenerational justice among past, present, and subsequent generations of creators, the prevailing intellectual property laws convert their traditional cultural bonds of building upon each other's advancement into licensor-licensee relationship. The innovations and creative expressions of subsequent creators will be stifled if they can not afford to pay for transformative, critical, derivative, parodistic, or recombinative uses of some ideas and information conveyed by pre-existing works (Carroll 2006, pp. 875–878; Nunzinato 2002, pp. 84–85; Oddi 2002, pp. 59–64; Rawls 1971, pp. 288–293). The reallocations of legal entitlements will be presumably obstructed if the economic interests of rights holders and future creators or their other motives are in conflict with each other. Since transformative, critical, derivative, or parodistic works frequently suppress demands for the original ones. The new intellectual property law regime discourages conversations among creators.

Since the law itself turns knowledge and information goods into commodities, it is difficult to employ precise utility calculus or price mechanism to measure whether the new and extended proprietary rights, their public legal enforcement and private ordering schemes can efficiently optimize their production, use, and dissemination in the global communications network; or to put it in other words, whether these public and private institutional arrangements are the most advantageous to enhancing social welfare of present and subsequent generations. The advocates of the prevailing intellectual property law regime usually refer to Demsetz' theory of the origin and efficiency of private property rights which is thought to buttress their strong arguments for the propertization of knowledge and information goods and the redefinition of property rights in digital environment (Epstein 2006, pp. 9–11; Frischmann 2006, pp. 12–14; Levmore 2003, pp. 192–94). Briefly, the argument is as follows: technological innovations make the bundle of previously determined property rights over knowledge and information goods incomplete because the new information and communications technologies provide new opportunities to create further economic benefits for non-owners without reaping the rewards of rights holders' investments. Externality problems and the allocative inefficiency of the market coordination apparently bring about the redefinition of the incomplete bundle of property rights, if the benefits of new property rights regime outweigh the costs of enforcement. In this view, the increase of the economic values of knowledge and information goods is corollary to the incessant process of privatization, the redefinition of existing property rights, and the private appropriation of their economic values in order to improve the allocative efficiency and smooth operation of the market. They simply leave out of consideration that the bundle of property rights over knowledge and information goods is always and inevitably incomplete because we do not possess perfect knowledge to predict and delineate all of their future uses and economic benefits in advance. In case of knowledge and information goods, incompleteness should be rather considered as the consequence of dynamic progress in the production, use, and dissemination of knowledge and information than of the imperfectly defined bundle of property rights. Besides its harmful effect, the permanent rearrangement of the bundle of property rights does not offer any constructive market solution to maximize private returns because it can not override the dynamic characteristic of knowledge and information production which generally increases their overall economic values. So, the use of knowledge and information goods always produces economic values for third-parties which come from their non-rival and non-excludable features as well as the positive feedback of the growth of knowledge in society. Briefly, the growth of knowledge and the flow of information in society are not uncompensated takings. Therefore, it is neither efficient nor constitutional if the law forces all recent and future beneficiaries into customer or licensee position in order to avoid externality problems. In spite of the profound differences of ownership between physical and symbolic assets, Demsetz became convinced after a slight hesitation in his influential essay that his land ownership paradigm is applicable to intellectual property rights system, too. He claims without further arguments that there are similar externality problems in the use of land and ideas. If externality problems emerge from the use of knowledge and information goods, the redefinition of property rights is necessary to restore the allocative efficiency of the market coordination. So, a well-defined proprietary rights regime automatically leads to the most efficient use of new ideas if the benefits derivable from these private assets will be concentrated on their creators (Demsetz 1967, p. 359). Following the line of Demsetzian analysis, the progressive propertization of knowledge and information goods and the parallel decrease of the public domain can be expected in order to maintain the efficient allocation of the market if the benefits of private rights holders' appropriation outweigh the costs of the enforcement of their legal entitlements.

As this misleading analogy between physical and symbolic assets illustrates, the new intellectual property rights regime is in lack of solid theoretical foundation. The laws justify the new provisions on the basis of natural rights theory and of utilitarian theory, respectively. Nevertheless, these two rival philosophical theories have different priorities. Rights-based approach does not allow any concession about the original rights and entitlements of rights holders on behalf of efficiency advantages while utilitarian approach does not rule out the possibility that the effort of welfare maximization may result in the denial of some rights and legal entitlements of rights holders. Giving priority to social benefits over the inherent rights of authors, the first intellectual property laws followed utilitarian approach. The prevailing intellectual property rights regime habitually claims the priority of the inherent rights and legal entitlements of rights holders over social welfare and leaves little room for other considerations as if it were the most beneficial arrangement for rights holders as well as society at large. In spite of the unique characteristics of the production, use, and dissemination of knowledge and information as they are explained above, the laws take the original rights and legal entitlements over them for granted. Although, the justifications of the inherent rights and initial entitlements over great many public domain materials – for instance, gene sequences, software algorithm, databases, news, press coverages, plain facts, words and expressions, obvious business methods, concepts, abstract ideas, and color schemes – to exclude others are vigorously disputed from legal, moral, and economic points of view (Levmore 2003, pp. 193–94; Merges 1999, pp. 581–588; Reichman and Uhlir 2003, pp. 319–322). On account of the economic benefits of exclusive rights and the absence of the legal protection of public interests, the fraudulent assertions of ownership notices to the public domain materials became widespread in the global communications network (Aufderheide and Jaszi 2004, 9–10; Mazzone 2006, pp. 1038–1047).

Whenever legislature assigns new property rights and valuable legal entitlements it also has to mimick the market, because the justification of the current rearrangement of property rights is based on presumed market failures. Though, knowledge and information goods do not behave like typical market goods in economic transactions. As a consequence of the redefinition of the bundle of intellectual property rights, knowledge and information goods as commodities rather resemble regulatory givings than real market goods: legislature provides exclusive and almost persistent ownership rights for individuals and corporate rights holders over them, sets the extended scope and lengths of their proprietary rights and legal entitlements, and regulates their production, use, and dissemination endorsed with technological measures and detailed statutory provisions (Dibadj 2003, pp. 1045–50; Liu 2004, pp. 87–166). Instead of price mechanism of competitive market or free bargaining over preferred terms and prices towards the equilibrium position of parties, prices and terms of use of knowledge and information goods are to a large extent determined by the original assignments of property rights and legal entitlements as well as monopoly power of rights holders. The rights holders' monopoly position creates market externalities, leads to endowment effect – namely, in property rights bargaining the owners value their legal entitlements higher and demand more money to exchange than they would be willing to pay for acquiring them –, and imposes significant social costs on society. The advocates of market solution have to puzzle out the following contradiction: meanwhile a simple market transaction may not efficiently allocate knowledge and information goods without the effective legal enforcement of rights holders' proprietary claims, a strict and comprehensive proprietary rights regime does not efficiently fulfill the purpose of maximizing their social values and creates disutilities.

The current changes in intellectual property laws regime result in losses of non-owners which they are usually averse to bear. Broad and widespread infringement of proprietary rights and legal entitlements is a case in point. From economic point of view, the infringements of owners' rights reduce the social costs of strict proprietary regime caused by exclusion, rent-seeking, fragmentation, endowment effect, and deadweight social losses, and to certain extent neutralize some of these detrimental effects on non-owners and society at large. These are almost unavoidable consequences of the current intellectual property laws because the rights holders' monopoly position and their legal enforcement prevent market mechanism or property rights bargaining to correct the inefficient allocations of property rights in digital environment. So, it is hard to prove that the incessant extension of legal entitlements and monopoly power of rights holders over determining prices and terms of use of knowledge and information goods in economic transaction maximize their social values and enhance social welfare. Contrary to Demsetz' insights into the efficiency of private ownership over the public domain, if the full benefits derivable from knowledge and information goods are concentrated on rights holders, it results their less efficient uses and diminishes their social values. In case of ideas, a lenient proprietary regime is likely to produce more utility for all parties and to create more welfare. Thus, economic theories over private property, efficiency, and incentives seem to offer a less feasible explanation of the motivation of recent trends in intellectual property legislation than public choice theory.

3. Private Ordering

The prevailing intellectual property laws, international treaties and agreements are captured by the vested interest of corporate rights holders in implementing strict and comprehensive proprietary rights over knowledge and information goods (Birnhack 2006, 516–517; Braithwaite and Drahos 2002; Elkin-Koren 2005; Frischmann 2006; Sell 2003). Nonetheless, corporate rights holders seldom stay within the bonds of the intellectual property laws and regularly use licenses and standard form contracts in order to contract around the provisions which they assume to be suboptimal for maximizing their own economic benefits. Briefly, they endeavor to minimize state intervention in order to seek more efficient allocation decisions on their behalf. In doing so, they supersede the nuanced rules of the current intellectual property laws and further expand their exclusive rights over knowledge and information goods – especially, plain facts, databases, public domain materials, words and expressions, mathematical formulae, compilations, color schemes, customer's personal data, obvious business methods, concepts, and so forth – to which they are initially not entitled by the law (Moffat 2007, pp. 11–20; Radin 2004, pp. 1–15; Lessig 2004). This trend somehow confirms Demsetz's prediction, namely, as the economic values of knowledge and information goods grow so further attempts on their private appropriation can be anticipated. The redefinition of property rights by the law and private ordering will inevitably occur if the benefits of new and enlarged property rights outweigh the costs of control and enforcement of legal and contractual obligations of non-owners.

The proliferation of private ordering by means of licenses and standard form contracts brings about the preemption of the provisions of the prevailing intellectual property laws. The prevalence of contractual arrangements in the global communications network would be welcome if it were relied upon the voluntary, rational, and deliberate consent of parties. Without doubt, fair contract provides the most efficient welfare-enhancing institution for coordinating the transfer and exchange of property rights and legal entitlements among the members of society. Alas, the great majority of contracts completed in the online world does not depend upon *aggregatio mentium* or *consensus ad idem*, namely, the meeting of the minds of the parties. Standard form contracts, license terms and conditions are deliberately drafted on a take-it or leave-it basis. Though, this general business practice is not devoid of paradox: private ordering regime offers an appealing perspective to wither away the public domain in order to maximize the economic benefits of corporate rights holders as well as to preempt the biased and restrictive provisions of the recent intellectual property laws by means of providing free – or at least less constrained – access to knowledge and information goods. Private ordering gives corporate rights holders further opportunities to create their own private legislation and to impose less favorable or even unfair and unreasonable terms and conditions on their non-drafting parties. At the same time, it could also fulfill the principle of distributive justice and equity among present and subsequent generations of creators and users as it is the case of many innovative initiatives like Free Software Foundation, Open Source Movement, Creative Commons, Open-CourseWare, iCommons, Science Commons, and so forth.

Whether private ordering regime serves the private interests of corporate rights holders or the public benefits solely depends on the inclination of rights holders. Their feeble inclination towards fair share principle is manifested in the fact that the great majority of rights holders prefers to enlarge their proprietary rights over the public domain materials and to create their own rights against the world rather than to offer fair and reasonable terms and conditions of the use of what they are entitled to own in order to "promote the progress of sciences and useful arts". In effect, private ordering could facilitate the spread of knowledge and the free flow of information, but typically takes the form of private legislation, which narrows the possibilities of non-contracting parties, as well. In this way, the matter of ownership becomes the key issue for virtually all kind of computer-mediated interactions to which exclusive rights are attached by the law or private legislation in the digital world. Therefore, private ordering could at most be a compassionate endeavor for protecting the public domain but not an effective panacea against the harmful economic, social, and cultural consequences of the prevailing intellectual property laws.

Regarding the production, use, and dissemination of knowledge and information as essentially cooperative enterprise, it is quite reserved proposal to appeal to rights holders for a gesture of goodwill, virtue, and decency without scrutinizing the legitimacy, social fairness, and economic efficiency of the current proprietary regime over knowledge and information goods as defined by the law. Well-defined and exclusive property rights are indispensable for promoting the allocative efficiency of the market in the world where economic resources are scarce, exhaustible, rival, and excludable. Though, the deference to this well-established paradigm about the primacy of private goods over public goods is neither self-evident nor particularly productive in case of knowledge and information to which these economic phenomena can be hardly applied.

References

Aufderheide, P. & Jaszi, P. (2004). *Untold Stories: Creative Consequences of the Rights Clearance Culture for Documentary Filmmakers.* Washington, D.C.: American University.

Balkin, J. M. (2004). Digital Speech and Democratic Culture: A Theory of Freedom of Expression for the Information Society. *New York University Law Review*, 79, Vol. 1, pp. 1–55.

Barthes, R. (1974). *S/Z.* New York: Hill and Wang.

Benkler, Y. (1999). *Constitutional Bounds of Database Protection: The Role of Judicial Review in the Creation and Definition of Private Rights in Information.* New York University School of Law, Public Law and Legal Theory Working Paper Series, WP 10.

Benkler, Y. (2006). *The Wealth of Networks: How Social Production Transforms Markets and Freedom.* New Haven, London: Yale University Press.

Birnhack, M. D. (2006). Global Copyright, Local Speech. *Cardozo Arts & Entertainment,* Vol. 24, pp. 491–547.

Boyle, J. (2000). The First Amendment and Cyberspace: The Clinton Years. *Law & Contemporary Problems,* 63, Winter/Spring, pp. 337–351.

Bradford, L. R. (2005). *Parody and Perception: Using Cognitive Research to Expand Fair Use in Copyright.* New York University, Law and Economics Research Paper Series, Working Paper No. 05–09.

Braithwaite, J. & Drahos, P. (2002). *Information Feudalism: Who Owns the Knowledge Economy?* London: Earthscan.

Carroll, M. W. (2006). One For All: The Problem of Uniformity Cost in Intellectual Property Law. *American University Law Review,* Vol. 55, pp. 845–900.

Demsetz, H. (1967). Towards a Theory of Property Rights. *The American Economic Review,* Vol. 57, 2, May, pp. 347–359.

Dibadj, R. (2003). Regulatory Givings and the Anticommons. *Ohio State Law Journal,* Vol. 64, No. 4, pp. 1041–1124.

Elkin-Koren, N. (2005). What Contracts Cannot Do: The Limits of Private Ordering in Facilitating a Creative Commons. *Fordham Law Review,* Vol. 74, No. 2, pp. 375–422.

Epstein, R. (1989). Justice across the Generations. *Texas Law Review,* 67, pp. 1465–1483.

Epstein, R. (2006). *The Structural Unity of Real and Intellectual Property. Progress on Point: Periodic Commentaries on the Policy Debate,* Release 13.

Gordon, Wendy J. (1992). On Owning Information: Intellectual Property and the Restitutionary Impulse, *Virginia Law Review,* 78, pp. 149–281.

Guibault, L. M.C.R. et al. (2007). *Study on the Implementation and Effect in Member States' Laws Of Directive 2001/29/EC.* Amsterdam, Institute for Information Law, University of Amsterdam.

Hardin, Garrett (1968): "The Tragedy of the Commons." *Science,* Vol. 162, 13 December. pp. 1243–1248.

Heald, P.J. (2007). *Property Rights and the Efficient Exploitation of Copyrighted Works: An Empirical Analysis of Public Domain and Copyrighted Fiction Best Sellers.* University of Georgia, School of Law Research Paper Series, No. 07-003.

Kaul, I., I. Grunberg, and M.A. Stern (1999). Defining Global Public Goods. In Kaul, I., I. Grunberg, and M.A. Stern, (Eds), *Global Public Goods: International Cooperation in the 21st Century.* Oxford: Oxford University Press, pp. 3–17.

Landes, W. M. & Posner, R. A. (2003). Indefinitely Renewable Copyright. *Univer-*

sity of Chicago Law Review, 70, pp. 471–488.

Lessig, L. (2004). *Free Culture: How Big Media Uses Technology and the Law to Lock Down Culture and Control Creativity.* New York: The Penguin Press.

Levmore, S. (2003). Property's Uneasy Path and Expanding Future. *University of Chicago Law Review*, Vol. 70, Winter, pp. 81–195.

Liu, J. P. (2004). Regulatory Copyright. *North Carolina Law Review,* Vol. 83, No. 1, pp. 87–166.

Mazzone, J. (2006). Copyfraud. *New York University Law Review,* Vol. 81, pp. 1026–1100.

Merges, R. P. (1999). As Many As Six Impossible Patents Before Breakfast: Property Rights for Business Concepts and Patent System Reform. *Berkeley Technology Law Journal,* Vol. 14, pp. 577–615.

Moffat, V. R. (2007). *Rethinking Contractual Restrictions on Fair Use: Preemption and the Structure of Copyright Policymaking.* University of Denver Sturm College of Law, Legal Research Paper Series, Working Paper No. 07–21

Netanel, N. W. (1996). Copyright and a Democratic Civil Society. *The Yale Law Journal*, 106, pp. 283–387.

Netanel, N. W. (2000). Market Hierarchy and Copyright in Our System of Free Expression. *Vanderbilt Law Review*, 53, November, pp. 1879–1932.

Nimmer, D. (2003). 'Fairest of them All' and Other Fairy Tales of Fair Use. *Law & Contemporary Problems,* 63 Winter/Spring, pp. 263–287.

Nunziato, D. C. (2002). Intergenerational Justice Between Authors In the Digital Age. *Journal of Intellectual Property Law*, Vol. 9, No. 2. pp. 219–290.

O'Rourke, M. A. (2000). Toward a Doctrine of Fair Use in Patent Law. *Columbia Law Review*, Vol. 100, No. 5, pp. 1177–1250.

Oddi, A. S. (2002). The Tragicomedy of the Public Domain in Intellectual Property Law. *Hastings Communications and Entertainment Law Journal,* Fall, pp. 1–64.

Posner, R. A. (1992). *Economic Analysis of Law*. Boston, Toronto, London: Little, Brown, and Company.

Radin, M. J. (2004). Regime Change in Intellectual Property: Superseding the Law of the State with the 'Law' of the Firm. *University of Ottawa Law and Technology Journal,* Vol. 1, Issue 1–2, pp. 173–188.

Rawls, J. (1971). *A Theory of Justice.* Cambridge: Harvard University Press.

Reichman, J. H. & Uhlir, P. F. (2003). A Contractually Reconstructed Research Commons for Scientific Data in a Highly Protectionist Intellectual Property Environment. *Law and Contemporary Problems,* Vol. 66, pp. 315–462.

Samuelson, P. A. (1954). The Pure Theory of Public Expenditure. *Review of Economics and Statics*, 36, November, pp. 387–389.

Schultz, J. (2002). The Myth of the 1976 Copyright "Chaos" Theory. http://www.lessig.org/blog/archives/jasonfinal.pdf last visited on 16 May 2007.

Sell, S. (2003). *Private Power, Public Law: The Globalization of Intellectual Property Rights.* Cambridge: Cambridge University Press.

Stiglitz, J. E. (1999). Knowledge as a Global Public Good. In I. Kaul, I. Grunberg, M. A. Stern (Eds), *Global Public Goods: International Cooperation in the 21st Century* (pp. 308–325). Oxford: Oxford University Press.

Vanneste, S., Van Hiel, A., Parisi, F., Depoorter, B. (2004). *From 'tragedy' to 'disaster': Welfare effects of commons and anticommons.* George Mason University, School of Law Law and Economics Working Paper Series, 04–23

Walker, J. (2003). The Digital Imprimatur: How Big Brother and Big Media Can Put the Internet Genie Back in the Bottle. *Knowledge, Technology, & Policy,* Vol. 16, No. 3, pp. 24–77.

James Robertson

Changing the Scoring System for the Game of Economic Life

The present mismatch between money values and ethical values is one reason for the growing interest in business ethics. Practical questions for discussion and study on business ethics should therefore include whether and how the mismatch can be rectified or at least reduced.

This paper outlines proposals that would significantly reduce it by reforming and developing the worldwide money system. It raises some philosophical and political questions, as well as institutional ones. It suggests that people concerned about the ethics of business should see the money system as the scoring system for the game of economic life and should consider how they could help to change it to get a better game.

The global financial collapse of recent months underlines the importance of these proposals. If they had been in force, they would have done much to prevent the collapse occuring and to limit the damage if anything like it had occured.

1. The Money System: A Scoring System that Needs to be Improved

It is now a commonplace of environmentalist thinking, if not yet of environmental practice, that preventing the creation of waste and pollution upstream at source may be more effective than treatments after the event "at the end of the pipe". But, in most spheres, higher priority is still given to remedial than to systemic ways of responding to problems – such as dealing with sickness rather than promoting healthy conditions of living, and responding to crime with penal solutions rather than by creating conditions that promote law-abiding behaviour.

Encouragingly, some of the contributions to the *2006 Interdisciplinary Yearbook of Business Ethics* discussed upstream approaches to the problems of business ethics. A significant cause of the problems is the mismatch between money values and ethical values. Business survival and success depend on making money, and will continue to do so. But, because the rewards and penalties offered by the money system are now often perverse, businesspeople find they have to compromise or suppress their ethical values in order to survive and succeed. So indeed does everyone else.

Few politicians, business leaders, public officials, economists, sociologists, political scientists and other professionals and academics have been interested in money as a whole system that might be made to work better. Most are preoccupied with achieving survival and success, accepting that the money system now works as it does. Even more noticeable is how few campaigners for ethical causes – social justice, a better deal for developing countries, ending poverty, fairer sharing of the world's resources, dealing with climate change, human rights and so on – seem to see the present system of money as an important cause of the ills they oppose. Why has its development over time left us with a system that many people experience as promoting economically inefficient, socially unjust, and environmentally disastrous outcomes?

Once we ask that question, the answer is clear. The primary interest of the goldsmiths and bankers and government servants who have evolved the monetary, banking and financial system over the centuries, and the primary interest of the great majority of the bankers and other financial specialists who manage it today, has been to make money for their customers, shareholders and other associates, and for themselves and their own organisations. Nobody has been responsible for seeing that the monetary and financial system works efficiently and fairly for all its users – locally, nationally and worldwide. Even in today's democratic societies virtually no economists and financial analysts and political scientists are employed to work out how it might be redesigned, evolved and managed from the standpoint of Adam Smith's "impartial spectator" (Buchan, 2006, p. 184) or John Rawls' "veil of ignorance" (Rawls 1972, p. 607).

The worldwide economy – at personal, household, local, regional, national, international and global levels – is obviously much more important than a game. But, imagining it as a game, we can think of money as its scoring system.[1] The scoring system for any game rewards some things and penalises others; and so, by motivating the players in how to play the game, it shapes what kind of game it is and what kind of outcomes it will have. Money's power in that respect is even greater than many other scoring systems; high enough money scores are needed to enable players to survive and succeed as the game goes on.

More and more people are now experiencing its effects as damaging and destructive. They see it as responsible:

- for systematically transferring wealth from poor people and countries to rich ones,
- for the money-must-grow imperative that compels people to make money in socially and environmentally damaging ways,
- for the diversion of economic effort and enterprise towards making money from money and from the rising values of existing assets, instead of from providing valuable goods and services,
- for systematic bias in favour of people, organisations and nations who should be managing the system efficiently and fairly on behalf of all,
- for eroding the credibility of political democracy, and
- for fuelling opposition to globalisation in its present form, and thereby threatening world peace and security.

The growing concern with business ethics is one of many signs that the workings of the money system now need to be changed.

So what needs to be done?

2. Money Cannot Be a Scientifically Objective Calculus of Value

The 20th century showed that a centralised socialist economy cannot work efficiently, justly or ecologically. But the idea that the only alternative is a free market economy based on prices dictated by objective money values outside human control is sheer fantasy. The way today's money system works reflects its historical development by people. It must surely be within the capacity of people today to develop it further now to reflect the needs of the 21st century.

Our starting point is to note that at the national level a government's principal operational functions include control of how the money supply is created and managed, control of how public revenue is raised by taxing and charging and borrowing, and control of how that public revenue is spent. These functions directly affect well between a third and a half of the total flow of money through modern economies. Prices and money values therefore reflect how these functions are carried out – including who creates new money and puts it into circulation, in what form; what is taxed or charged for, at what rates; and what government spends its money on, in what amounts. In other words, however the government carries out those functions it will inevitably skew the price structure of the whole economy against some things and in favour of others. The proverbial 'level playing field' is simply a mirage.

This means that the democratic state has a crucially important responsibility which at present goes by default. It should manage its own money dealings in ways that provide a framework that encourages its citizens and their organisations to deal with *their* money in ways that automatically tend to serve the interests of other people while serving their own. The "invisible hand" of supply and demand should be programmed to serve the interests of all participants in the economy.

In that context the teaching attributed to Milton Friedman that there ain't no such thing as a free lunch (TANSTAAFL) should be seen as nonsense. By profiting from more than their fair share of the value of common resources, powerful individuals and organisations and nations now enjoy massive free lunches. Examples of common resources are given later. As inputs to essential economic activity their value should provide a source of public revenue, not private profit.

Before going more deeply into the monetary and financial responsibilities of governments, we should note that global and local money problems cannot be effectively handled at national or Eurozone level. "The Global Dimension" is discussed below. Local monetary and financial initiatives, like LETSystems, time banks and community development banks, have been growing in numbers in many countries to support local economic exchange and trading. They can usefully complement the mainstream money system and should be encouraged.

3. The Monetary and Financial Responsibilities of the National State

The state, then has three main operational monetary and financial tasks:

1) to arrange that a sufficient *money supply* of official currency (pound, euro, dollar, etc) circulates efficiently and fairly;
2) to collect *public revenue* for spending on public purposes; and
3) to administer *public spending* programs.[2]

3.1 The Money Supply

In a democratic society one might expect that all the money created as additions to the national money supply backed by the state would be created by an agency of the state, that it would be spent into circulation on public purposes, and that it would be created debt-free.

What actually happens, however, is quite different.[3] In the UK, for example, less than 5% of today's national money supply is created debt-free by the Bank of England and the Royal Mint as banknotes and coins. Over 95% is created by commercial banks writing it electronically into their customers' bank accounts out of thin air as profit-making loans. It has been estimated that UK commercial banks make over £20 billion a year in interest from creating this electronic bank-account money, whereas the issue of banknotes and coins brings in public revenue of less than £3bn a year. It has also been estimated that additional public revenue of about £45bn a year would result from a reform that

- prohibited commercial banks from creating bank account money, as they are prohibited from creating banknotes and coin, and
- gave the Bank of England the task of creating it and passing it as debt-free public revenue to the government to spend into circulation.

Treasury ministers have so far been unwilling to ask their experts in the Treasury or the Bank if those financial estimates are broadly correct.

The following are among the arguments for a reform on those lines.

1) The official-currency money supply is a public resource. The value of creating it should be a source of public revenue. To allow it to be captured as private profit is both economically damaging and unfair to particular sections of society.

2) Creating money as interest-bearing debt is systematically inflationary. A debt-based money supply means that more money than has been created is always needed to pay back the debt involved in its creation – not only the "principal" (the sum borrowed) but also the interest payments on it while it is outstanding. That is why the main objective of monetary policy now has to be a target level of inflation – within a bracket of, say, 2–3% a year.

3) Creating money as debt is pro-cyclical. It tends to amplify the volatility of the business cycle instead of damping it, because banks want to lend more and bank customers want to borrow more when the economy is booming, and less when it is depressed. It thus contradicts the anticyclical aim of monetary policy.

4) If the great majority of new money entering circulation is channelled into the investment and spending priorities of commercial banks and their customers, it creates economic distortion in favour of speculative investment in the value of existing assets, and against

productive investment to produce new goods and services. For example it encourages speculative investment in land – one reason for the spectacular rise in house prices compared with other prices and wages and salaries in many countries in recent years.

5) One inevitable feature of an economy in which money is almost entirely created as debt is greater total indebtedness – higher levels of debt for citizens, companies and government. This causes:
   a) Social damage and injustice. It artificially widens the gap between poor and rich. It increases the flows of money from poor to rich, since the poor have greater need to borrow money and the rich are better placed to lend it.
   b) Environmental damage and destruction. When the money needed for all transactions has to include a sum to pay the interest arising from the way it was created, organisations and individuals are compelled to convert natural resources into money more rapidly than they might otherwise need to.

6) Allowing commercial banks the privilege of creating, free out of thin air, the main resource they need as input to their business gives them a subsidy enjoyed by virtually no other industry. It protects the big, established commercial banks from competition from smaller, more enterprising, efficient and customer-friendly potential new entrants to their various lines of business, including the country's main system for making and receiving payments. This reduces the economic efficiency both of the wider financial services industry and of the economy as a whole.

7) How money is now created and what the effects of that are, should not remain veiled in mystery. Widespread failure to understand how the monetary and financial system now works is a serious impediment to its improvement.

Official monetary statistics recognise that electronic money in current bank accounts constitutes by far the biggest proportion of today's money supply. Nonetheless, supposedly knowledgeable opponents of reform argue that it isn't really money, it's only "credit". Unwittingly they alert us to a parallel with the past. In the 19th century, when paper banknotes were finally recognised to be real money, the Bank of England became the only bank in England allowed to issue them. Its banknotes still say "I promise to pay..." – a reminder that banknotes originated centuries earlier as credit notes from private businesses, but everyone now knows they are now cash, not just credit notes. Similarly, almost everyone recognises that electronic money in current bank accounts has become real money immediately available for spending. The continuing creation of state-backed electronic money for private-sector profit by commercial banks is now a glaring anachronism.

Prime Minister Gordon Brown, when UK Chancellor of the Exchequer (Minister of Finance), took a significant preliminary step in 1997 towards bringing monetary control up-to-date. He gave the Bank of England, which had been nationalised in 1946, operational independence to manage monetary policy in accordance with the government's published policy objectives. But the Bank can still only influence how much new money the commercial banks create, by regulating interest rates, thus the price of borrowing, thus bank customers' demand for loans, and thus the amount of new money the banks create. It is time now for the obvious next step – to make the Bank clearly responsible on behalf of the state for actually creating the required amounts of state-backed electronic bank-account money, just as state agencies create new banknotes and coins.

3.2 Public Revenue

Existing taxes are becoming less viable. For example:

- National economies in a *competitive global economy* have to reduce taxes on incomes, profits and capital to attract investment capital and highly qualified people – both being increasingly mobile.
- *Ageing societies* will be unable to support growing numbers of "economically inactive" people by taxing the work and enterprise of fewer people of working age.
- *Internet trading* is making it more difficult for governments to collect customs duties, value added tax and other taxes and levies on sales,

and easier for companies and rich individuals to shift earnings and profits to low-tax regimes and tax havens.

- *Tax avoidance* by big corporations and rich individuals is reaching crisis proportions. Estimates are that tax havens cost £255bn annually to governments worldwide, and hold assets of $11.5 trillion ($11,500bn), causing serious distortion of economic priorities and supporting criminal money laundering (Tax Justice Network 2005).

Shifting a large part of the tax burden on to the value of land and other common resources which cannot be moved abroad will probably be national governments' most effective response to these problems.

As well as becoming less viable, existing patterns of taxation are now positively perverse:

- By heavily taxing employment and rewards for work and enterprise and lightly taxing the use of common resources, they systematically encourage inefficiency in all kinds of resource use – under-use and under-development of human resources, and over-use of natural resources (including energy and the environment's capacity to absorb pollution, including carbon emissions); and
- By taxing the value added by most people's positive contributions to society (VAT), and failing to tax value subtracted by those who make most profit from common resources, they systematically skew the overall burden of tax in favour of a rich minority.

These facts argue, on both economic and ethical grounds, for a "tax shift" on to the use of, or profits from, the value of common resources.

Common resources are resources whose value is due to nature and to the activities and demands of society as a whole, and not to the efforts or skill of individual people or organisations. The site value of land is the most obvious example. The value of a particular land-site, excluding the value of what has been built on it, is almost wholly due to the activities and plans of society around it. For example, in the UK when the route of the London Underground Jubilee Line was published, properties along the route jumped in value. Access to them was going to be much improved. A public policy decision and subsequent investment of public money, gave owners of those properties a £13bn windfall financial gain.

They had done nothing for it; they had paid nothing for it; they had been given a very large free lunch. By contrast, the UK Treasury raised £22.5bn for UK taxpayers in 2000 by auctioning twenty-year licences to use the radio spectrum for the third generation of mobile phones. The governments of other European countries also raised significant sums that way.

Local government in the UK, including the Greater London Authority, has been exploring whether rail and road transport developments could be financed out of the increases in property values which they generate. In 2004, the Vice Chair of Transport for London summarised some of the arguments for land value taxation (Wetzel, 2004):

> "With income from LVT … the government could provide new public transport infrastructure; abolish economically damaging property taxes such as council tax, business rates and stamp duty; raise personal allowances so that millions of lower-paid workers pay no income tax at all; and reduce VAT rates to help consumers and businesses. The tax would improve earned incomes; cut the cost of tax collection; provide affordable homes; reduce urban sprawl; avoid property-led business booms and slumps; and minimise the need for constant changes in interest rates to control land prices".

Pressures for a shift to environmental taxation have recently risen along with awareness of global warming and other threats too – including world shortages of energy, food and drinkable water, and worsening pollution of the oceans.

In addition to land-sites, the electro-magnetic spectrum, the national money supply, and the environment's capacity to absorb pollution and wastes, important common resources include: the value of unextracted energy; limited space available for road traffic, airport landing slots, etc; and water for extraction, for fishing and for waterborne transport. The annual value of these and other common resources is very great, and rises along with the world's economic growth.

3.3 Public Spending

The overall structure of public spending programs needs more searching scrutiny by politicians, the media and the public than it now gets. Two examples illustrate this.

First, $1.5 to $2 trillion a year is estimated to be spent worldwide on *perverse subsidies* which encourage economically, socially and environmentally damaging activities (Myers, 1998). These include the subsidies from rich-country governments to their farming and agricultural sectors, which – combined with tariffs against imported food – devastate those sectors in poorer countries and expose the hypocrisy of rich-country support for free trade. But there are many other examples of perverse subsidies. Sustained national and international determination is needed to reduce them year by year.

Second, support for a *basic income* (or Citizen's Income) continues to grow, especially in Europe but elsewhere too.[4] It would be paid to all citizens as of right, out of public revenue. It would include state pensions and child allowances, it would replace many other existing social benefits, and it would eliminate almost all tax allowances, tax reliefs and tax credits. It would recognise that, in a society of responsible citizens, some of the public revenue arising from the value of common resources should be shared directly among them. Politicians and government officials now channel huge sums in contracts and subsidies to private-sector business and finance, as well as to governmental organisations, to provide citizens with public services. Much of that public money could be given directly to citizens to spend for themselves in a market economy made more responsive to their needs by the other reforms proposed. It would especially help poor people who would not benefit from reductions in income tax but would have to pay the new environmental taxes.

3.4 Summary – The Responsibilities of the State

The state should carry out its three main operational monetary and financial responsibilities in ways that will distribute the value of common resources among all citizens and reduce or even abolish taxes on earnings and profits from providing useful goods and services. This will create a new framework of prices which reward the market economy for delivering outcomes which combine economic efficiency with social justice and environmental care. The state will then be able to let the market economy operate more freely, with less intervention, than now.

People in business will then find not only that their businesses need to be less closely regulated by the state but that they themselves are troubled less by conflicts between the ethical values held by themselves, their families and their friends and the money values that must necessarily shape their business behaviour if they are to survive and succeed.

Business people and everyone else too, as citizens, will experience greater freedom at the personal level. A Citizen's Income will allow them, if they wish to do so, to reduce the amount of money they must earn by working as employees. Then, with more time and energy to supply themselves and their families with some of the goods and services they now have to buy, they will be able to further reduce their need to spend money if they want to. As consumers, employees and savers they will be in a stronger position to influence – and choose between – the people they have to deal with in those capacities.

4. The Global Dimension

Particularly acute problems of business ethics tend to arise in the international sphere. This is at least partly because the international institutions that deal with world monetary management, public revenue and public spending, are out of date. As at the national level, their further development should be based on sharing the value of common resources more fairly.

Over ten years ago the independent international Commission on Global Governance recognised the urgent need for international monetary reform in a globalised world economy (Commission on Global Governance, 1995). Since then there has been growing criticism of the present 'dollar hegemony' of the United States. For using the dollar as the main global currency, the rest of the world is estimated to pay the US at least $400bn a year. A Pentagon analyst justified this as payment to the US for keeping world order; others see it as a way for the richest country in the world to compel poorer ones to pay for its unsustainable consumption of global resources (Douthwaite, 2002). To build up their reserves, poor countries have to borrow dollars from the US at interest rates as high as 18% and lend it back to the US in the form of Treasury

Bonds at 3% (Romilly Greenhill and Ann Pettifor 2002). The dollar is a global monetary instrument that the US, and only the US, can produce; world trade is now "a game in which the US produces dollars and the rest of the world produces things that dollars can buy" (Lin 2002).

A genuine international currency, issued by a world monetary authority, is needed as an alternative to the US dollar and other 'reserve currencies' like the yen, euro and pound. Issuing it would provide a source of revenue to the world community, as national monetary reform would provide a source of revenue for nation states.

The Commission on Global Governance also recognised the need for global taxation "to service the needs of the global neighbourhood". It proposed making nations pay for use of global commons, including:

- Ocean fishing, sea-bed mining, sea lanes, flight lanes, outer space, and the electro-magnetic spectrum; and for
- Activities that pollute and damage the global environment, or cause hazards beyond national boundaries, such as emissions of CO_2 and CFCs, oil spills, and dumping wastes at sea.

Revenue from global money creation and global taxes would then provide stable sources of finance for global spending by organisations like the United Nations, including international peace-keeping programmes. Some of it should be distributed per capita to national governments, reflecting the right of every person in the world to a global 'Citizen's Income' as a share in the value of global resources.

This approach:

- Would encourage environmentally sustainable development world-wide;
- It would generate a much needed source of revenue for the UN;
- It would provide substantial financial transfers to developing countries by right and without strings, as compensation for rich countries' disproportionate use of world resources;
- It would help to liberate developing countries from dependence on grants and loans from rich-country-dominated institutions like the World Bank and International Monetary Fund;

- It would help to solve the problem of Third World debt;
- It would recognise the shared status of all people as citizens of the world; and
- By helping to reduce the spreading sense of injustice in a globalised world, it would contribute to global security.

5. A New Direction of Economic Development

The changes in the money system outlined here will help to create a new direction of economic development.

Attention will shift to creating well-being for people and the Earth; to enabling people to develop their capability, rather than reinforcing their dependency; and to conserving the Earth, rather than transforming its resources as rapidly as possible into money. The fairer sharing of the value of common resources will help to decentralise power and wealth – both by giving a fairer deal to people in their own places and by requiring rich and powerful people and corporations and nations to bear their full share of the environmental and social costs of centralisation. The new framework of monetary and financial incentives will automatically harness self-interest to common interest within and between nations.

Reducing the present mismatch between money values and ethical values will reduce the need to swim against the financial tide in order to act in accordance with their ethical values. Increasing numbers of business leaders and other business people may, therefore, decide that the proposed reforms of the money system deserve their support.

Supporting them need not imply a commitment to any particular ideology. They offer the prospect of

1) the democratic state performing its monetary and financial functions more purposefully and effectively, thereby being able
2) to allow the market economy to operate more freely.

They will also make it financially easier

3) for people to reduce their present degree of dependence on goods and services and jobs provided by big corporations and the state, and

4) for both people and organisations to act in ways that conserve, not squander, natural resources.

Those interested in ideologies could interpret the first item as part of a socialist program, the second as part of a capitalist program, and the third and fourth as parts of a program for people and Earth that helps to modernise both socialism and capitalism.

Notes

1 The rules of the game – laws, regulations, treaties and other features of legal and judicial systems – are, of course, also important. Company law, defining the rights of different stakeholders in businesses, is directly relevant to business ethics. But changes in the legal obligations of companies will be unrealistic if they conflict with what the money system rewards and penalises.

2 The state also regulates non-state financial institutions.

3 The facts and figures in this section apply to the UK (Huber & Robertson, 2000; Robertson and Bunzl, 2003). But the case for reform applies to other countries too.

4 Basic Income Earth Network is one of the most useful among the growing numbers of websites in this field – http://www.etes.ucl.ac.be/BIEN/Index.html.

References

Buchan, J. (2006). *Adam Smith and the pursuit of perfect liberty.* London: Profile Books.

Commisssion on Global Governance (1995). *Our Global Neighbourhood.* Oxford: Oxford University Press.

Douthwaite, R. (2002). *Defense and the Dollar*, Dublin, FEASTA (Foundation for the Economics of Sustainability). www.feasta.org.

Greenhill, R. & Pettifor, A. (2002). *The United States as a HIPC (Heavily Indebted Prosperous Country) – How the Poor Are Financing the Rich.* London: New Economics Foundation. www.neweconomics.org.

Huber, J. & Robertson, J. (2000). *Creating New Money – A Monetary Reform for the Information Age.* London: New Economics Foundation. www.neweconomics.org.

Lin, H. (2002). *US Dollar Hegemony Has Got To Go*. Asia Times Online, www.atimes.com/global-econ/DD11Dj01.html.

Myers, N. (1998). *Perverse Subsidies: Tax $s Undercutting Our Economies and Environments Alike*. Winnipeg, Canada: International Institute for Sustainable Development.

Rawls, J. (1972). *A Theory of Justice*. Oxford: Clarendon.

Robertson, J. & Bunzl, J. (2003). *Monetary Reform – Making it Happen*. London: International Simultaneous Policy Organisation. www.simpol.org.

Tax Justice Network (2005). *Tax Us If You Can.* London, www.taxjustice.net/cms/front_content.php?idcat=30 ("understand how tax havens, capital flight and tax evasion are harming global society and economic wellbeing").

Wetzel, D. (2004). *The Case for Taxing Land.* London: *New Statesman*, September 20.

Frans de Clerck
Triodos Bank Group

Ethical Banking*

This paper introduces "ethical, sustainable, social, alternative, development or solidarity" banking and finance. While how they are described differs, these financial institutions have much in common. They share a values-driven approach to their business, at odds with the conventional commercial banking and finance industry. The reader will find an overview of the pioneers in this field, the historical context they worked in, and the initiatives they created. In particular, this paper explores the case of Triodos Bank from an ethical perspective.

1. Introduction

The private sector and financial institutions in particular, are increasingly expected to play an important role in helping to create a truly sustainable world. Conscious consumers, ethical investors, enlightened

* Inspired by insights and publications from Wilhelm-Ernst Barkhoff, Rolf Kerler, Paul Mackay, Peter Blom, Lex Bos, Henri Nouyrit, Christine Gruwez and collegues from the ethical banking scene.

businesses, non-governmental organizations (NGOs), cultural creatives, and leading international institutions are working to make the triple bottom line (People, Planet and Profit) a reality. These people and institutions want solutions for poverty, injustice, war, widespread diseases, educational inequalities, destruction of nature and the planet. In short, they want to improve quality of life for everyone on this planet, and they understand that we all are economically interdependent and share responsibility for taking care of current and future generations. They share a common set of values centered on the enhancement of human dignity. Some of these values are already universally accepted, others are nurtured at a personal level. These people and institutions know that knowledge, capital and energy are available to solve most problems, and they see themselves as part of the solution. They feel, and are, ready to contribute.

Non-governmental organizations are calling upon financial institutions to implement more socially and environmentally responsible lending policies. An example of this is *The Collevecchio Declaration on Financial Institutions and Sustainability*[1], which outlines six principles that financial institutions should embrace: a commitment to sustainability, 'do no harm', responsibility, accountability, transparency and sustainable markets and governance. In addition it explicitly calls on financial institutions to advocate for regulation.

Some developments are making their way into mainstream decision-making in the financial industry. Clear indications of this are demonstrated by major financial institutions signing internationally recognized standards such as *The Principles for Responsible Investment*[2], (which concern environmental, social, and corporate governance issues), and *The Equator Principles*[3]. And although it is a long way to change policies in the banking sector, especially when the driver is the 'common good' and an ethical approach, some first steps have been made. This mainly concerns the creation of ethical investment funds, separate from the core banking business and the introduction of lending policies with ecological criteria for the sake of risk reduction or reputation protection. However, ethical private development banking as a core mission is still not on the agenda of institutions mainly focused on financial return, driven by stock markets and short-sighted shareholders.

Crucial human development and sustainability issues, such as poverty alleviation, awareness of the human impact on climate change, investments in renewable energy, fair trade and organic agriculture, demonstrate possibilities for a widening of the scope of responsibility for financial institutions. But three conditions need to be fulfilled by financial institutions. They have to:

1) understand and accept their responsibility in handling money-streams
2) establish internal rules and systems that can cope with ethical criteria
3) widen their concept of return including the social and ecological added-value of their activities.

As this is probably more than one step too far for most financial institutions, the near future of truly sustainable ethical banking will be furthered by specialized banking institutions with ethics as part of their DNA and mission, and with financial activities clearly oriented towards human development, quality of life and a decent financial profit.

2. Ethics in Banking

Banking and finance as a profession have an intrinsic value chain which is interwoven with the cycle of providing adequate financial products and services. As long as there are no bank guidelines or criteria on ethical, social and sustainability aspects, the individual co-worker or the lending committee generally apply the 'neutrality rule', excluding ethical, social and environmental considerations from bankers′ decision making. In reality however, money is not neutral and it involves responsibilities from its inception and along the distribution chain where it has to do with value creation, not only pure financial value but also human, social and environmental added values.

Money, capital, intelligently and wisely invested as an instrument for improving quality of life, can have a major impact on human development. Because of this impact, a neutral attitude to investment and lending is irresponsible. In the financial markets, money and money systems become mechanical and develop uncontrollable dynamics. Financial regulators and authorities are only concerned with the mechanics of the system in order to prevent major breakdowns. Is there an organizational design for money as an instrument subservient to human development? What are ethical impulses and human qualities that can be found in modern societies in both developed and developing countries and that can be brought into the banking and finance process? Three possible impulses are described below.

- *The impulse of brotherhood and sisterhood at interpersonal, local and global level*

  Are we interested in each other's physical existence and well-being? Do we feel responsible for each other? How do we deal with this question on a planetary level? In which way do we experience and organize a global co-existence at a time of different development patterns in different cultures and in different natural environments around the globe? Are we ready for such a scope of social cohesion while self-interest and pure consumer orientation are taking the lead in modern economy? Can a transparent money stream serve social cohesion and stimulate the interest in each other by making money become available to those who are talented to use it in value creation activities for the common good?

- *The impulse of recognition of human dignity as a precondition for human development*

  The impulse of recognition of human dignity as a precondition for human development demonstrates a deep interest in the personality and the capacities of other human being(s), including respect for a person's inner life and active tolerance. How can the availability of money, in its respective qualities through lending, investment or donation, contribute to a valuable use of human capacities?

- *The impulse of searching for 'meaning' and 'quality' in life*

  'Meaning' refers to a constant quest for understanding, including the spiritual level. 'Quality' has to do with the added value that is the outcome of a search process where choices are being made in life. How can investment and lending be directed to meaningful positive action and be diverted from financing negative developments or negative aspects of an undertaking? Can ethical banking be a method of constant search and reflection on the meaning of human and economic value creation while putting its findings into practice?

Standing in the middle of social and economic developments, bankers are well positioned to have an overview and a feeling for what matters, although they assess risk versus opportunities without considering social and environmental development. They generally use this position to grow their business. They do not take this opportunity to transform the knowledge they have acquired into wisdom that they could apply in developing ethical banking policies and making fundamental choices.

Bankers' observations of the needs of their clients and of society in general can lead to inner reflection and understanding of the degree of importance of some development questions. Conscious bankers can transform feelings of powerlessness into an understanding that something can be done. Transparency of ethical banking operations – showing what is financed – is a prerequisite for open dialogue with clients and civil society. This dialogue can lead to a deepening of understanding of the phenomena and to inspiration for adequate action to be deployed. When this perpetual process of observation, reflection, mutual exchange, taking responsibility, action and reporting is included in specific organizational forms, ethically working bankers will have developed a valuable instrument that is not only serving the needs of their clients but will also help to fulfill the needs of society as a whole. This description of ethical banking does not refer to charitable action. It starts from the observation that altruism, or looking after someone else, is part of economic life where division of labor and interdependency of people are a basic principle of efficiency. Human needs are an expression of a healthy egoism in an economic process dealing with the fulfillment of needs. Altruism in an economic sense is not in contradiction with egoism but tends to equilibrate the economic process.

3. Emergence of Ethical Banking and Finance

Quite early in history gold, reflecting the spiritual world, served artistic, religious and economic goals, and was directly linked with the gods and their servants, the priests, who organized its flow. Throughout medieval times Christianity set its laws on usury, Islam set its rules on interest, and monasteries organized economic life in their surroundings, working with investments and charitable actions in a moral and religious perspective. In these times humanity was strongly organized around three realities: the spiritual world, the world of nature, and local social entities.

Since the beginning of the 15th Century, natural sciences and later enlightenment, gradually emancipated people from the world of the gods, nature and their local social environment. The relationship between human beings changed with the growing predominance of individualism.

This context and background of modern society are fruitful to the emergence of modern ethical banking concepts and practices.

3.1 Essential Characteristics of Banking on Values

Social, ethical, alternative, sustainable, development and solidarity banking and finance are denominations that are currently used to express particular ways of working with money, based on non-financial deliberations. A precise and unified definition of these types of finance as such is not available and perhaps not possible because of the different traditions from which ethical finance actors have emerged. While individual motivations from founders, investors, savers, borrowers, social entrepreneurs, managers and co-workers of these institutions vary greatly, there are some universal human values, practices and needs that motivate all of them to develop positive action. Conscious handling of money is considered to be an additional value in itself. Many of these values are part of internationally recognized declarations or principles, such as the United Nations *Universal Declaration of Human Rights* (1948) and the International Labour Organization's *Declaration on Fundamental Principles and Rights at Work* (1998), that identify basic rights such as:

- Freedom of thought, opinion and expression using reason and conscience are leading to financing art and culture, education and research
- Equal rights at a political and juridical level, the freedom and right of association in a democratic society and the right to work are a basis for financing civil society projects and for participating in the public debate about the benefits and challenges of shared social responsibility
- A spirit of brotherhood, based on understanding, tolerance and cooperation in economic life leads to financing social entrepreneurs especially in the areas of high urgency like poverty alleviation, fair trade, environmental production and preservation.

To practitioners of ethical banking, raising consciousness and responsibility are essential in their missions and ambitions. They make the choice to only finance projects and organizations that contribute to a more sustainable society and they define absolute criteria about who they will not lend money to, for example non-sustainable products and/or services and those involving unsustainable working or production processes. Their specific products and services reflect these values and intentions.

While money is a catch-word of our age, to ethical banking institutions and their shareholders, savers, investors and borrowers money and ethical banking practices are instruments for human development. These characteristics differ with those of mainstream finance, mainly driven by market forces, shareholder value and financial return.

3.2 Socially Responsible Investment

In the 18th century, the Quakers in the United Kingdom refrained from investing in industries they were morally opposed such as tobacco, alcohol, gambling and the slave trade. This was the first negative ethical screening of investments, later to become known as Social Responsible Investment (SRI). It continued into the 1920's with the Methodist

Church of North America screening out negative activities, or 'sin stocks' from their investment portfolios. In the 1960's and 70's the conviction that investment funds could be used to achieve social change give rise to the public demand for ethical investment vehicles such as the Pax World Fund[4]. In the 1980's investments supporting the South African apartheid regime were avoided, and Friends Provident (UK) was the first financial institution to launch an SRI fund. With its help, the Ethical Investment Research Service (EIRIS) was established to provide critical research and information on stock-listed companies' social, environmental and ethical performance. In the United States, Amy Domini developed her ethical screening advice services and the first ethical stock market index.

At the beginning of the 1990's, a first attempt was made in The Netherlands to develop a positive ethical screening to be used alongside the original negative ones. This positive screening involves a best-in-class method, where company performances were compared with those of competitors. This type of screening has since further been developed and several ethical screening organizations have been established. Standards of screening have been developed and screening services are now being widely provided to banks, insurance companies, asset managers, private bankers, institutions and high net-worth individuals. Most stock-listed companies have had some form of ethical screening of their social and ecological behavior so that ethical funds or asset managers can constitute diversified portfolios primarily based on combined negative and positive ethical criteria. Some of these funds, such as those of the Triodos Bank Group are also actively involved in (proxy) voting at shareholder meetings. The ethical investment fund market is developing quickly and many mainstream banks are offering such products.

Today there are more than 600 ethical investment funds worldwide and their number is constantly increasing. However the ethical quality of these products differs significantly in terms of quantity and content of positive and negative criteria applied. As a quality label the generic denomination ethical fund, indicating that some sort of ethical screening has been applied, is not appropriate.

As corporations have a tremendous impact on both people and planet, and as they are operating more globally than ever, their corporate responsibility needs to be engaged. It is making its way to the boardroom table as well as that of management and has begun to become integrated into internal structures. However high-quality corporate responsibility is still an exception. Whether responding to customers' needs, preparing and positioning for the future, or as a result of enlightened leadership, this development is likely to grow and so will the number of ethical questions and dilemmas. By applying ethical screening to their investments, ethical funds, institutional investors, and pension funds are exercising influence on management, and gradually corporations are responding with improved transparency, reporting and accountability.

In the best circumstances ethical screening and investor pressure is contributing to a process of intensified observation, questioning, reflection, measurement, ethically amended business principles and consequently adapted decision-making. Better reporting, external social and environmental auditing, the elaboration of social and environmental guidelines in corporate governance codes, feedback by the screening analysts and regulations could lead to a system of permanent upgrading of ethical conduct by corporations.

Socially responsible investment is of a totally different nature than ethical banking since it relates to the ability to influence company behavior through the provision of capital to stock-listed companies. Ethical banking, as described below essentially relates to direct financing and loans.

3.3 Ethical Banking

Ethical banking provides direct finance through lending and risk capital to fulfill the financial needs of selected entrepreneurs, organizations and businesses. The cooperative movement from the beginning of the 20th Century is an example of how essential needs can be fulfilled through forms of collaboration and mutuality in membership organizations. Modern forms of cooperation beyond focusing on membership needs such as the fair trade and microfinance movements, combining economic with social values, are a step forward in the understanding and practice of brotherhood and solidarity in a global economical context. Both the cooperative movement and the new social movements from the 1960's have developed a practice of ethical banking. Cooperative banks and new social banks co-exist, while some mainstream banks have become aware of business opportunities in this sector. Microfinance institutions focus their effort in parts of the world where there is a high need for poverty alleviation. For a better understanding it is useful to distinguish these tendencies:

Cooperative banks and credit unions

Cooperative banks and credit unions have substantially contributed to the provision of finance to their members, which at the beginning of the 20th Century was a social task. This changed when commercial and savings banks started offering banking services on a broader scale. Many cooperative banks expanded their activities into the mainstream and lost their special social mission. Some of them have recently rediscovered their roots and are redirecting some of their activities. Driven by a need to build a specific brand identity in a financial world where there is much of the same, these banks manage to successfully combine usual banking business (the bulk of their financial operations) with support to specific areas such as community development, the not-for-profit sector and/or environmental development. Examples of such banks include Rabobank in The Netherlands (having a major green fund), Vancity in Canada (giving low-income and marginalized members access to necessary financial services), Cooperative Bank in the United Kingdom (taking a stand against the finance of armaments), and Crédit Coopératif in France (developing solidarity products).

New social banks or private development banks

In the last 40 years, new social banks or private development banks have been created and new banks are still being constituted. Impulses for their mission come from the recognition of social and human development needs and of a need for quality of life including care for the environment. They look to the processes of dealing with money, not only at the outcome. They see cooperation not as a mutual aid process between members but as a shift of interest towards the needs of other human beings in a local or global context. They want to stay true to their values even as they grow and change, while growth is not a target on its own and financial profitability is seen as a condition for further development.

These impulses are connected to those driving non-governmental organizations such as Amnesty International, Greenpeace and Friends of the Earth, and they appeal to those citizens or cultural creatives who are convinced that they can play an active role in this global and personal development process.

The founders of GLS Bank in Germany, constituted in 1974, were the first to concentrate on the qualities of loan money (to potentially stimulate human interest) and gift money (the most productive seed capital). They also focused on the capacity building force of bringing savers and borrowers, consumers and entrepreneurs together for investment, for example in organic agriculture, school education and care for handicapped people. GLS sees banking as a continuous and conscious process of directing the money flow to where it is needed in societal and human development perspective. Individual responsibility and care for the other human beings are seen as core drivers of these processes. Community building through participation in these processes is stimulated through the creation of borrowing and guarantor communities, dedicated savings instruments, and a choice for clients of the bank to determine for themselves the height of interest rates on their deposits. This ethical approach to banking has been an inspiration for many of the European social banks which have gradually developed over the last few decades. Notwithstanding cultural differences, variety in size, accents (social, environmental), products and services, and stage of development, all of them have ethical and sustainable development elements at the core of their mission, ambitions and practices. All of them are making a good case for human and social development while offering both generally and specifically designed products and services to their respective markets. Whilst a few have failed, most of them have found their way of continuity, with different models of functioning, whilst being in conformity with general banking regulations. An overview of those successful institutions that have a banking statute are:

- ShoreBank, (1973), USA
- GLS Bank, (1974), Germany
- Triodos Bank, (1980), The Netherlands with branches in Belgium, United Kingdom, Spain and Germany
- Freie Gemeinschaftsbank in der Schweiz, (1984), Switzerland
- Merkur Bank, (1985), Denmark
- Wainwright Bank and Trust Cy, (1987), USA
- Alternative Bank Schweiz, (1990), Switzerland
- Cultura Sparebank, (1997), Norway
- Ekobanken, (1998), Sweden
- Banca Popolare Etica, (1998), Italy
- Charity Bank, (2002), United Kingdom.

Some social banks have been constituted by trade unions and have developed based on social and ethical criteria:

- ASN Bank (1960), The Netherlands
- Caisse d´Economie solidaire Desjardins (1971), Canada
- Unity Trust Bank (1984), United Kingdom.

Other social banks are focusing on some specific market segments:

- Health and social economy – Bank für Sozialwirtschaft, created in 1923 on behalf of the UN High Commissioner for Refugees and originally serving as the central administration for UN funding in Germany
- Environment – Umweltbank (1995) in Germany, and the New Resource Bank (2006) in the USA

These banks are quite different as to the volume of their operations – balance sheet totals vary from EUR 30 million to several billions, and their financing capacity from EUR 50,000 to EUR 25 million per project. All together they are currently financing tens of thousands of projects with added social value mobilizing the savings of more than one million people and institutions. Being still relatively small, these banking on values institutions, have substantial growth rates, are professionalizing and consider themselves to be catalysts for social change. With these banks also succeeding in applying outstanding internal organization and staffing practices, and in developing specific methodologies to properly deal with the ethical aspects of this type, they have a potential for further qualitative development.

Microfinance banks

Microfinance is a methodology of banking for the unbankables (people without access to finance), contributing to poverty alleviation through micro lending for income generating activities of the poor themselves. Although this methodology is not new nor comparable to the movement that was launched in Germany by Friedrich Wilhelm Raiffeisen at the end of the 19th Century (and later developed by credit unions), microfinance in its present form received a tremendous boost from the Grameen approach in Bangladesh, designed by the 2006 Nobel Prize winner Professor Muhammad Yunus.

In 2007 there were approximately 10,000 microfinance institutions worldwide. Apart from their contribution to economic development of millions of poor entrepreneurs, their families and their communities, they are often providing basic education and methods of community building. Some of these institutions have the potential to develop into full social banks and are helped with support structures from the north such as Oikocredit, launched in 1975 by the World Council of Churches, the Triodos Microfinance Funds (1994 and 2002), ShoreBank and Shorecap International (1988 and 2003) and many other institutions. Apart from Grameen Bank some of the most advanced microfinance institutions are Brac Bank (Bangladesh and Afghanistan), Basix (India), Acleda Bank (Cambodia), Mibanco (Peru), Findesa (Nicaragua), Compartamos (Mexico), Equity Bank and K-Rep Bank (Kenya) and Centenary Bank (Uganda).

The financing of poor people's entrepreneurship in the north requires different methods compared to traditional banking due to the different social structures and the predominance of individualism. Adie (Association pour le droit à l´initiative économique) created in 1989 by Maria Nowak in France, is a good example of collaboration between mainstream banks, government and non-governmental organizations.

So long as microfinance institutions are able to integrate basic ethical values, going beyond the mission of fighting poverty, and are able to connect local savings to local borrowing and continue to get the support from northern development money, they have potential for high quality development. New challenges however, such as the effects of climate change, especially in the south, will require huge investments from the world community pointing at the necessity of further social and environmental globalization on the planet.

4. Innovative Practices Case Study: Triodos Bank Group

The paradigm shift of people using banks to take their money out of the anonymity of markets and to give it a human sense needs drastically innovative banking practices. The Triodos Bank Group is interesting because of its constant transformation, growth, diversification and integration while keeping its mission and ambitions alive. This is why today it is a reference for the social, ethical and sustainable banking industry.

Using money as an instrument for social change can bring a breath of fresh air to the banking sector. This is what the founders of Triodos Bank in the 1970's had in mind. After more than 25 years of social banking activities, and with more than EUR 3 billion assets under management with profits linked to ideals, the Triodos Bank Group has widely diversified its activities beyond taking in savings and lending to sustainable projects and social entrepreneurs. Apart from traditional payment services, the bank's core activities include lending to sustainable projects and social entrepreneurs, taking in savings from committed depositors, special purpose fund management, ethical screening services, and sustainable private banking. Some of these and the innovative processes of human resource management and internal organization are described in more detail below to demonstrate how an ethical approach to banking can be successfully integrated in business and organizational processes.

4.1 Lending and Investment

The Triodos Bank, apart from the funds under management, currently lends more than EUR 1 billion to more than 6000 projects, organizations and businesses in four key areas and 16 sub-sectors in five European countries. The key areas are nature and environment, social business and culture and society and sub-sectors include organic food, renewable energy, social entrepreneurship, housing associations, education, art and culture. Through investment and lending to 77 microfinance institutions in 35 countries (Asia, Africa and South-America) and financing fair trade and development cooperation, the Triodos Bank Group is contributing globally. The Group also manages environmental, social and ethical investment funds.

The lending policy integrates both positive criteria (projects that combine added cultural, social or environmental value with financial credibility and that clearly benefit the wider community) and negative criteria (clearly defined non-sustainable products and services, for example, nuclear energy, environmentally hazardous substances and the weapons industry plus non-sustainable work processes, such as intensive agricultural production, genetic engineering, breach of fundamental labor rights).

The following elements of policy and practice concerning lending are innovations with ethical considerations and demonstrate the quality of internal ethical and social standards:

- the Bank's loan book is 100% mission based (ethical, social and development aspects in the mission) which is high compared to lending by other social banks and exceptional compared to mainstream banking practices where lending is ethically neutral
- choices of lending are clearly defined and are only in selected key areas and sub-sectors. The lending process ensures that each selected project meets absolute criteria which measure the potential negative impact of a borrower's activity on people and the environment

- lending criteria are strict, publicly available and in accordance with the Bank's other activities such as risk capital investments, investment funds under management and other asset management activities. The criteria are regularly reviewed and refined as changes in the market and trends develop
- the degree of transparency on lending activities is very high due to systematic publications on the lending activities
- the bank closely considers the motivations of the people involved in a loan application as much as their trustworthiness.

4.2 Savings, Advice and Screening Criteria

By practicing a high degree of transparency – savers are told what the Bank is financing with their money – clients are stimulated to develop their interest and participation in positive action deployed by social entrepreneurs and other borrowers at the bank. Donations in the cultural and development sphere are systematically stimulated and Private Banking clients receive personal advice on the ethical aspects of their investments. The screening criteria used by the Triodos Bank Group's ethical funds are among the most strict and severe in the ethical funds sector. Ethical screening services and advice are also made available to other banks, pension funds and institutional investors.

4.3 Human Resource Management

The Triodos Bank pays special attention to the social and ethical motivation of co-workers throughout recruitment, evaluation procedures, internal training programs and debate and exchange sessions. Co-workers are stimulated to take initiatives out of their personal conviction and ethical understanding as far as these are in line with the Bank's mission. Co-workers are encouraged to actively participate in the decision making process. Therefore information, discussion, reflection, and creative inputs at weekly co-workers meetings are a tradition that keeps the attention, also on ethical questions, alive. The salary system employed by Triodos Bank is based on the principle that income is generated by the joint efforts of all co-workers. Salaries are set based on a system of job evaluation, and on a self-imposed restriction of the difference between the lowest and the highest salaries at the bank.

4.4 Internal Organization

The main objective of Triodos Bank is of an ethical nature – 'With the exercising of banking business, the company aims to contribute to social renewal, based on the principle that every human being should be able to develop in freedom, has equal rights and is responsible for the consequences of his economic actions for fellow human beings and for the earth'[5].

The voting rights at the Triodos Bank annual general meeting are exercised by the Foundation for the Administration of Triodos Bank Shares (SAAT), and are guided by the ethical goals of the bank thus preserving its identity.

Triodos Bank's annual report[6] has been formulated according to the Global Reporting Initiative (GRI) guidelines since 2001 when it was the first bank worldwide to publish an integrated annual report (social, environmental and economic). Since then, more than 50 banks worldwide have used the GRI guidelines[7] for their sustainability reporting.

For reasons of principle, no share option scheme is offered to members of the Board of Management, Supervisory Board members or members of SAAT's Board of Trustees. The Group is structured in an integrated way, both locally and internationally, with business units covering specific activities. This helps the Group to consider the ethical aspects of the business at each level of business and globally.

5. Conclusions

Although private community and development banks, microfinance banks, ethical, environmental and social banks and ethical funds differ in terms of focus, accents, clients, products and business culture, they have in common to practice banking and investment with a human development mission. The differences tend to be rather complementary qualities that can be fertile in combination with each other. They are all delivering an innovative and human value contribution to the value-neutral financial system. Ethical banking as it has been described above stands in a historical line of continuous search for the application of ethical principles in banking and is in line with broader trends in the 20th and 21st centuries such as the emergence of civil society and the new social class of cultural creatives, growing consumer awareness, social justice and environmental movements and the growing recognition of social entrepreneurship, to name a few.

Some questions require continuous attention: Will this emerging financial business sector be able to achieve the relative scale and the professionalism to challenge the dominance of mainstream finance? Will the exceptions of the financial industry become the exceptional and a factor in modern society? Will a profound way of dealing with ethical choices be overruled by the superficiality of business development – also in ethical banking? Can ethical banking as a process with an instrumental character avoid becoming institutionalized? Can ethical banking be a portal for trust forces, morality and responsibility to feed money processes and the financial system with basic values and practices that can counteract uncontrollable growth?

Ethics is more than ever a subject of personal choice, behavior and responsibility. At the same time, more and more people are looking for values to incorporate in daily life. As contemporaries on their way, they are part of an ongoing process of search and practice linking up and networking with other people, creating new forms of social cohesion. Instruments such as ethical banking processes, products and services and money as a subservient tool can be helpful.

This paper is not intended to provide an in-depth analysis and screening of ethical banking practices. Instead, it describes an emerging financial sector, believing that it could become a significant factor in society, less in terms of volume and more through the human added value it delivers.

Notes

1 *The Collevecchio Declaration on Financial Institutions and Sustainability* was launched at the World Economic Forum in 2003.

2 *The Principles for Responsible Investment* is an initiative of the UNEP Finance Initiative and the UN Global Compact: see www.unpri.org.

3 *The Equator Principles* are a banking industry framework for addressing environmental and social risks in project financing: see www.equator-principles.com.

4 Inspired by a civilian wanting to invest in a mutual fund that did not invest in war-related industries, Pax World Fund was established in 1971 creating the first broadly diversified, publicly available mutual fund to use social as well as financial criteria in its investment decisions: see www.paxworld.com.

5 Triodos Bank Articles of Association, Article 2.

6 Triodos Bank's website: see www.triodos.com.

7 The Global Reporting Initiative is a multi-stakeholder process and independent institution whose mission is to develop and disseminate globally applicable sustainability reporting guidelines: see www.globalreporting.org.

PART 3

Opinions: The Challenge of Future Generations

Zsolt Boda
Associate Editor

Opinions

Our unsustainable economy, the depletion of natural resources, the destruction of eco-systems and the long-range effects of modern technologies, such as nuclear energy or genetic modification, obviously affect the quality of life of future generations. However, the rights and interests of yet unborn people are not represented in the decision making processes and institutional setting of today's society.

What kind of reforms could be possibly proposed so that the regulatory institutions of the global economy represent the rights and interests of the coming generations? What can business do in order to consider its effects on future generations? Are there principles, norms and decision making rules to follow? Or is the respect for the future is a challenge still to be met by business ethics thinking?

Jakob von Uexkull
Founder, World Future Council

Representing Future Generations

The interests of future generations need institutional representation, either by an organisation with moral power such as the Councils of Seers Into The Future in pre-colonial India or the World Future Council (WFC) today (see www.worldfuturecouncil.org). The objective for creating the WFC has been: creating a trusted, long-term voice to speak up for future generations; building an integrated global forum based on our shared ethical values; and overcoming action gaps by linking moral authority with democratic power. The Council unites fifty highly respected figures from across the globe and from all walks of life to create a strong ethical voice to represent future generations. Selected through a global consultation process involving 2,500 civil society organisations, the Council members bring with them a wealth of experience and expertise. Institutions that defend the *status quo* are well established, yet the *future* has no comparable voice. The WFC is being set up to close this gap. It will help create a fair and 'future-proof' world, by accelerating major changes in the direction of progress such as assuring a sustainable ecological base for human development, and fairness in the conduct of world affairs. Alternatively, there could be Future Councils as constitutional bodies with the legal mandate ('future veto') to ensure that we fulfil our responsibilities to treat the world as on loan and not as private property.

The business of business should be service to society, not preserving privileges and maximizing profits. Clearly, profitability is necessary to cover costs and living expenses but that is very different from maximizing profits by externalizing costs at the expense of our children. Let us take the example of climate change. We must stabilize climate chaos through a rapid acceleration of the transition to clean, secure, renewable and decentralized energy for all, while reducing the global demand of energy in total. The WFC also defined climate stabilization as fundamental necessity and human responsibility for a more just, peaceful and sustainable world for future generations. Clearly, business could and should actively contribute to this objective. Businesses active in areas which entail substantial future risks should be legally changed from limited liability Corporations to Foundations mandated to act for the common good.

Despite unprecedented knowledge, skills and resources, humanity is today on a collision course with its own future. Global resource depletion and pollution, growing inequalities between rich and poor, and the resulting potential for violent conflict are causing fear and uncertainty. Decisions taken today have longer-term impacts than ever before, yet short-termism has become a dominant trend. But there are and have always been common ethical rules, e.g. to treat others as we expect to be treated (a rule common to all faiths) and to hand over an improved or at least not a deteriorated planet to our children (common to all societies, otherwise humanity would be long extinct!). In market economies, markets are servants. Today we have market societies where markets masters with terrible consequences. But we can decide to follow the rules of ethics rather than the rules of profit.

Benedek Javor
Protect the Future (Vedegylet),
Budapest

A Speechless Mass Behind Sustainability

If, as Ernest Partridge states "philosophers are those troublesome individuals who ask the next question"[1], there is always a "next question" to pose by philosophers concerning the widely used conception of sustainability. This question is: why should we make efforts to form a sustainable economic and social system, when most of the benefits of these efforts will not be harvested by those individuals, who make them and who bear the expenses? Without having an appropriate and convincing reply to this question the society and the economic sphere can not be persuaded to give up some of its possible gains in favor of the obscure and uncertain conception of sustainability. As human and social rights are a widely accepted and secure basis for incorporating social costs into the economic activity (taxes, labor rights, collective contracts, etc.), rights of future generations are the final argument behind the claim for a sustainable system. Unborn people should be let in under the protecting umbrella of basic human rights.

To manage the moral obligations, which burden our collective actions, politics itself, we need tools in harmony with the logic of collective action. To make the moral imperative which is created by the rights of future generations lead not to a guilty conscience but to concrete actions, it must be translated to the code of social activity that is the language of law and politics. We need such legal-institutional solutions, which are able to appropriately perform the representation of long-term interests in the intricate system of social and political decision-making.

We could say that these interests and the institutional or legal solutions to defend them exist to exactly such an extent, which is minimally sufficient to their just and fair representation. Environmental regulations, sustainable development strategies, development plans, institutions to control financial equilibrium of the states or the possibilities which are ensured for the civil society to represent basic social values outside the classical political decision-making processes are on the whole suitable to set the delicate balance between the interest of future and present generations. Theoretically we could say, that present structures and institutions are able to fulfill these tasks, and there is no need to rack our brains on new institutional or legal solutions.

Nevertheless, practice confutes this optimistic assumption. All the surveys, monitoring the state of our natural environment show the continuous degradation of the global and the European ecosystems (e.g. Living Planet Report 2008, European Environmental Agency Report 2005, etc.), instead of emission decrease, targeted in the Kyoto Protocol, most of the EU member states has produced a significant (some states more than 20%) augmentation in the emission of greenhouse gases, and the EU as whole also does not meet with its burden-sharing targets. The continent is fighting with a more and more serious demographic crisis, and connected to this, with unsustainable pension systems, the rate of state indebtedness is mostly increasing. The revision of the Sustainable Development Strategy of the EU, which was accepted in Gothenburg in 2001, clearly shows, that the Union has to face with grave failures in the implementation of its own strategy and targets. This is indicative of the disability of present structures in effectively enforcing the principles of sustainability and intergenerational justice, and proves that the welfare of our present societies is assured increasingly at the expense of the future generations.

In this situation, as our Hungarian NGO, Vedegylet-Protect the Future! concluded, without new institutional tools this trend can not be turned back. In 2000, we initiated to set up the office of the Ombudsman for Future Generations in Hungary. The bill prepared by Mr. Laszlo Solyom, the present president of Hungary, and member of the board of Vedegylet-Protect the Future! at that time, proposes the establishment of an institution that, while striving to interpret the unsure and "soft" outlines of the protection of the succeeding generations as widely as possible, sets the reference base of its interventions in the effective Hungarian legal system so that the Commissioner's recommendations and procedures be built on a „hard" legal base so their validity cannot be challenged. Under this dual pressure we outlined an institution possessing extremely strong competences in some fields as compared to the ombudsmen's usual competence, in other cases its task stands closer to the "advocate" ombudsman's role or the "guardian's" status described well in international legal literature. This way the Ombudsman gets effective tools in his hands, legal guns which, besides having an influence on the conscience and goodwill of the society and the decision makers, establish the possibility of enforcing legal remedy in the case of decisions negatively affecting future generations – including environmental conflicts as well as decisions concerning pension systems, state debts, infrastructural investments, which burden unjustly future humanity.

In spite of our efforts, the bill was not accepted by the Parliament for years. Although we had to wait for the final success of the proposal, the issue is held in the public discussion for more than seven years, and we had the hope to see this institution in work one day. However, till the bill was not accepted, we initiated a special program, the Representation of Future Generations (REFUGE), to act like a 'civic ombudsman', to fulfill those tasks which originally were delegated to the planned ombudsman office. By the end, this work brought about a remarkable success. After seven years, as a consequence of renewed lobby activity and negotiations with parliamentary parties, an all-party agreement was formulated around our proposal. Months of reconciliation and hard debates on the exact text ended in November 2007 a public, all-party introduction of the bill on the Ombudsman for Future Generations. Soon, the Hungarian Parliament accepted the bill, creating an internationally unique, model-like institution for the protection of the interests of future generations. And in May 2008 also the new ombudsman was finally elected. We believe, that his appearance in the decision-making processes really makes a difference, and helps Hungary to ensure a more livable natural and human environment for future generations.

But there is also an other important issue. State level solutions can be essential, but with the deepening European integration process we have to count with the fact, that an increasing proportion of decision-making competencies and also of the responsibility towards succeeding generations is concentrated in the supranational structures of the European Union. We can not neglect this shift of decision-making and power centrums either in the protection of future generations. In 2005 we started an initiative for setting up the European representation of future generations. With preparing a proposal which consists of different possible solutions, we aimed to initiate a lively professional and political debate on the institutional protection of future generations in the EU. We are convinced, that without institutional replies to the problem, all strategies and plans on sustainability stay mere papers.

Of course, there are problems with the representation of future generations. We cannot have exact knowledge about the needs and necessities of yet unborn people in the future. However, some of their basic interests, like fundamental living conditions, healthy environment, etc. can be make probable. For the protection of these essential interests we need an institution which is not a decision-making centre (not to overemphasize uncertain 'rights'), but is able to effectively emerge and enforce these ethical considerations into our present decision-making processes (not to totally deprive future humanity of the possibility to raise their voices). Otherwise the silent majority, future generations will unavoidably fall victim of our present game.

Note

1 Partridge, E: *Environmental Ethics: An Introduction*, http://gadfly.igc.org/e-ethics/Intro-ee.htm

Paula Tiihonen
Committee Counsel to the
Committee for the Future,
The Parliament of Finland

The Right of Future Generations

In all social decisions – in politics and business – the future must always be taken seriously into consideration. But to be able to do this, you need forums for futures-oriented thinking and above all forums to make political, financial and juridical decisions on the future. These forums must be situated as close to centres of power as possible.

In Finland the Eduskunta, – which as the national parliament is certainly one of the centres of power – decided in 1993 to establish a special committee to deliberate problems of our common future, and thus also the rights and responsibilities of future generations. It was tentatively named the Committee of the Future, but in retrospect I would say it got off to a very good start in that at its very first meeting the 17 members (all parliamentarians, representing the full spectrum of political parties) decided to change the name to the Committee *for* the Future. The change of preposition demonstrated that they wanted their committee to be active rather than passive; to be for the future not against it, to take an innovative attitude to society, science and technology and not resist it merely because it is new, and so on.

In 1906 Finland was the first country in the world to give full voting and candidature rights to women in politics, which can be an explanation for why, after 100 years of this kind of great social innovation, the same Finnish Parliament was the first in the world to decide that our common future is so important that politicians also have to take real responsibility for it.

It is an adage of political life at any level that the first step to power is to take the initiative and put yourself in a position where you can set the agenda. In the Eduskunta, the Committee for the Future has taken this adage seriously from the very beginning. The only rule in setting an agenda has been that it has to be something that is new and important to people. Of course the idea is to tackle only big issues, but we have to be humble and admit that we see small things better. Some of them can turn out to be big matters.

Having a mission and working hard during these 13 years, the committee has taken its place in the Finnish parliamentary system as an innovative political body. Over the years it has created a new forum that works at the core of the parliamentary system and – even more important – has demonstrated that parliamentary measures can still be used to take the initiative within democracy.

The powers of the Eduskunta are as in any other Parliament. They can be divided into legislative, budgetary and visionary. The Committee for the Future was created to strengthen the visionary aspect. When the rights of future generations are the subject of discussion, it is specifically visionary power that is at issue.

Politics in this context is about values, attitudes, atmosphere and opinion building, and, most important, opinion leading. Nowadays politics is against change and against new things. It is said, and even believed, that voters do not want new things. But in this democratic system of ours the point is just the opposite, which is why politics is needed: to support new ideas, and among them, new innovations, both social and technical ones. Even when it is a question of really fundamental institutions, which you certainly should preserve and protect, there is need to discuss connections to other institutions, which are changing for sure. Even with the oldest institutions of mankind, the family, you have to be ready to understand new ways of thinking and new ways of doing, both now and in the future. The environment in which people are living as family members is always on the move.

Following and understanding the changing world is not enough. In order to be committed and effective politicians have to be active. They have to be creative actors and take a positive initiative. Governments and Parliaments are generally quite tradition-bound, so is this an impossible task? No. But you need new, innovative institutions that create new forums to think and talk about the challenges of the future. A similar remedy can be proposed for other social institutions, like businesses, as well.

Laura Nash
Cambridge, Massachusetts

The Legacy of Business Leaders

A discussion of the rights of the unrepresented future generations leads me back to the present: in as much as we accept the duty to universalize general rights today, we must extend those rights to others as yet unknown and unborn. I believe this is a matter of logic but also one of the highest expressions of human nature. To paraphrase former Harvard president A. Lawrence Lowell, humankind "was made for infinite conceptions."[1] The extension of our moral vision to the right of future generations is one of those exercises in infinite conception that characterizes the uniquely human capacity to transcend the boundaries of time, knowledge and physicality.

We have become increasingly aware that the conditions for guaranteeing environmental and economic sustainability in the future will most likely involve a scarcity of natural resources and political and scientific complexity on a scale we have not seen before. Before quibbling about the theoretical content of those rights, it is critical to establish a legacy of framing perspectives on present and future rights that solves for viability in executing our obligations. Ethical norms are best supported by the twin pillars of reason (logic, consistency) and world view (theory of human nature and sentiments of human relationship to the social and physical environment). Future generations will depend on our ability to morph our natural sentiments that favor growth over decay into something other than limitless consumption. Equally important, we will need imaginative leadership to frame these sentiments of sustainability through some other change agenda than "What must we give up?"

I summarize this transformation of worldview in the term, "a reasoned sense of Just Enough"[2] in contrast to our Never Enough world in which progress and the good life are bound to the pursuit of an infinite "More" while calls for cutting back on consumption (of food, fuel, land, or investmentable world wealth) are associated with an unpleasantly constrained life for the sake of environmental viability. The latter leads politicians to the emotionally futile strategy of pursuing unpleasant solutions for the sake of life itself.

The business leader that can articulate and execute a strategy of Just Enough on environmental responsibilities will create a legacy most conducive to securing the rights of future generations. Getting there depends on the cultivation of more energizing assumptions that sheer cutbacks, and a special kind of emotional intelligence about future generations. It begins with a teleological bias that environmental responsibility should be approached through strategies of opportunity rather than strategies of deprivation for business itself, its shareholders, and consumers who use its products. Human nature seems to favor this approach, innovation requires it.

Secondly, as leaders and governing boards deliberate an ethic of natural resources sustainability, their framing question should move beyond conservation to integrate a full articulation of conception of the good life, concept that would include generativity. The political agenda must be not just, "how do we change to ensure that natural resources *last*, but also, *how do we create a platform of energy use (technology, consumption, process for allocating access) to ensure that the next generations can succeed in ways we feel are important to the human condition?* Businesses adopting this approach will tie principles of stewardship and sustainability to our deepest values

By taking an additive, positive approach, crisis solutions for deprivation can be abandoned in favor of an opportunistic pursuit of the good life as represented by the emotional experience of Just Enough (experienced satisfaction). This perspective is less about "stop" and more about "extend". Extend our time frame, extend our property rights connections beyond national borders, extend our emotions of caring to a broader, as yet even unseen population. Extend our capacity for satisfaction by deepening our ability to experience multiple and new forms of satisfaction rather than simply seeking ways of getting more of the same. Instead of more highways to temporarily increase driving satisfaction, why not more forms of workplace locations for today's corporation, ones that increase social interaction and decrease automobile dependency? Such extensions of mind and heart must inform our energy strategies and economic policies for business. Businesses are particularly good at trying to provide people what they want in order to profit. By seeing environmentalism as an opportunity to provide people with a future they want, they will extend their competitive imagination to, say, not only creating attractive hybrid autos that let people drive *more* on less fuel but creating products and services that improve workspace alternatives so that driving more is not the only pleasure and necessity under consideration.

These framing perspectives cannot be applied without vision. Specifically, business leaders must not shy away from a careful articulation of *the way they feel the world works when it is working right, and how particular business plans fit into that vision of a world beyond the legal entity that is the corporation and its balance sheet.* Skip that step and no amount of transactional ethical rule-making will sustain the future in ways worth the doing.

In our research for our book, *Just Enough*, Howard Stevenson and I found that even extremely selfish human natures seem to have a deep need for self-perpetuation, or generativity. The world working right is a world in which something positive of ourselves is left behind when we are gone our values, our ethics, our aesthetics, our ability to thrive. We labeled this drive "Legacy" in our study of business leaders who sustained multiple forms of success in their lives. It is a good model for understanding our legacy for ensuring rights to future generations, but typically the idea of legacy is not approached this way. For many, environmental legacy is about hair-shirts and bland food. Religious worldviews frequently frame Legacy as something you get rather than give: it becomes more about paving the way for your life in the hereafter (frequently with self-denial in the present) than paving the way for others to succeed. And yet the most universal moral and religious principle, the Golden Rule, is only made actionable in terms of the future success of others.

We observed a surprising and satisfying emotional payback in the way some leaders approached Legacy. For them, Legacy concerns motivate the active creation of a platform to carry forward that which is important (values) *in a way that helps others succeed in the future.* China becomes not just a source of carbon cap tradeoffs but an environment in which we and they have the opportunity to prosper from environmental innovations made now. As we wade through the uncertainties of global cooperation in the marketplace over energy use, business perspective must include a Legacy vision of the success of future generations how will they thrive, serve each other, exercise freedom?

This is a far more complex idea than the typical Legacy question one hears in business and politics: 'what will they remember us for'? That question plays to a different Legacy drive the desire to live forever, to control everything for all time, or to operate by a winner-take-all ethos and get yours now. As we are discovering in our attempt to cut back on fuel consumption, addressing the future from such narrowly defined self-interest is hard to moderate, however enlightened we are about the science of the climate. It feeds on a sense of Never Enough so that environmentalism becomes a subtractive exercise in our lives. Our moral intentions to leave the world in a sustainable state environmentally and in terms of ethical rights demands that we develop the capacity to understand there is a positive in Just Enough. That Never Enough burns people out, or fails to balance competing interests a must in any business.

Ironically, we will need to extend the idea of just enough to environmentalism itself, and resist totalitarian environmental solutions from either the right or the left. The recent all-out embrace of corn-based ethanol is a cautionary example: we may have unwittingly unleashed even more environmental waste and hunger in our haste to cut back on oil. Scarily we roll along, and in our necessarily imperfect forecasting, businesses can quickly wonder "why bother?", ducking responsibility through sheer loss of confidence or sheer passivity. With problems on the scale of global warming, what can one company do, anyway? Responsibility to the next generation, then, will rest first on an emotional maturity to stay the course not just on adrenaline and scare scenarios that motivate a quick fix, but on a reasoned sense that experimentation, and the adoption of new forms of consumption that lead to new satisfactions in life style are better than win-or-lose all.

I don't think it is easy for people to commit to what they know is, inevitably, only a partial solution. If you personally change your consumption habits, if your business invests in sustainable energy practices and product development, and that provides an example to others, great. But it won't change CO_2 levels today. Like saving one starfish on the beach in spite of losing so many more, these seemingly "futile" commitments need the emotional support of knowing Just Enough at a deeply ethical and emotional level. Religious and secular exercises in mindfulness help cultivate such felt satisfaction, and with it the patience and maturity to think through the complex technological, economic, and social challenges that environmental sustainability poses. The businesses that lead in cultivating such emotions and perspective will create a legacy worth banking on. Using these suggested perspectives, businesses can not only face the idea that they have an obligation to factor in the rights of future generations, they can create a viable approach to doing so.

Notes

1 The source of this widely cited quote is unknown.

2 Laura Nash and Howard Stevenson, *Just Enough. Tools for Creating Success in Work and Life,"* John Wiley Press 2004.

J. M. Sampath
Arpitha Associates
Bangalore

Enhancing the Quality of Our Decisions for Nurturing a Sustainable World

What can we do to consider our legacy for future generations? An effective way would be to deliberate on how we make our decisions. To help make decisions we have evolved strategic thinking, lateral thinking, critical thinking, systems thinking to name a few. Each geared to enhance our decision making process. Yet our collective decisions have left us in doubt about our collective futures. We seem not to have evolved a fundamental process, which will aid us in making decisions that would sustain the future.

Over decades corporate leadership unconsciously or subconsciously has lost its balance between vision and values in the decisions they took and gave in to the basic instincts of greed for wealth and power. As a result we are paying a huge price and in turn wondering why, what, when, where, who and how of building a sustainable world.

A look at the cover stories of the magazine Fortune or what is being taught in the leading business schools of the world points out to how everything boils down to how to increase the shareholder's value. Unfortunately society doesn't figure in the shareholder's list, neither does environment. Who figures is the investor and the only focus is on how to increase the returns i.e. create more and more return on investments

(ROI). Those who could do it got amply rewarded. People in power and those making things happen closed their eyes to compromising values and processes in the effort to increase profitability. The result was Enron, WorldCom and so on. It was a clear wake up call. But have we really woken up? We made the boards and companies more accountable on one hand yet on the other we increased the pressure on the CEO to give a better ROI, quarter on quarter, than his predecessor. The CEO uses the collective intelligence in the organization to find ways to beat the systems and processes to show a better ROI quarter on quarter. This pressure had forced DELL to compromise and when exposed, it had to restate four years of financial results, reducing net income by over $50 million for each year. This is a good example of how we keep beating the system. While making their students intelligent beings, our leading business schools are not necessarily teaching them the wisdom of knowing what to use their intelligence for. The outcome is evident.

Over the years with or without our own knowledge we have disintegrated the fundamentals of long term sustainable success and have operated at sheer execution level.

It is important to revisit the basic questions behind every business and the various levels at which it should be aligned and then take a look at every decision from a wholesome perspective so that we don't land ourselves into a further mess.

1. Basis for the Foundation

In order to understand the foundation that supports the execution of complex business decisions which will meet and surpass the expectations of all stakeholders including society at large, the following are critical questions every leader needs to address at the very conception of the business.

1) What do we want to achieve?

2) How do we want to achieve?

3) Why of what & how?

4. When to do what?

Each of these questions will have to be seen at different levels for us to get a wholesome perspective while running a business and contributing through it to build a sustainable world.

Let us examine each of these questions and the various levels at which they need a response and perceive how they enable leaders to deliver results.

2. Foundational Questions at Different Levels

The first critical question a leader in any business should ask is "What do we want to achieve?" This question when answered at different levels would give myriad insights into the business and its modus operandi.

At a fundamental level the answer to this question should give the vision of the business. It helps one to set the direction. And why must one examine the fundamental level while making a choice on setting the direction of a business? This is because at a fundamental level any venture cannot hurt the earth or the environment for in the long run it will lead to death of the business itself. Another important component of the fundamental level is that you cannot make profits at the cost of someone or something else, for this will deplete the resources in the long run and hence hurt the business. In other words someone's loss cannot be the foundation for another's gain. On the contrary everyone's gain should be the foundation of someone's gain. This would result in long term sustainable growth of the business.

While the above are few of the important elements that will show up when you explore what you want to achieve at a fundamental level, there will be many more elements that will open up which will lay a strong foundation for sustainable business and universal well-being.

The same question at a business life cycle level will give insights into the purpose of the business. Many businesses may pursue same direction but may have different purposes. For instance Toyota and Volvo may be in the same business, while Toyota may focus on efficiency, Volvo focuses on safety. Once the purpose is clear every time a conflict arises it becomes easy to move on since the purpose is clear. Many a time when the purpose behind the vision is neither clear nor established the focus of the business gets lost and a kind of disintegration begins. Some of the elementary natures of purpose are, it has to enhance the vision and bring into focus what is being attempted in the business.

What we want to achieve at an execution level will enable setting goals or milestones which allow the business to operate. The goals will have to be in line with the purpose. When the goals deviate from the purpose the business loses sight of the vision and its sustainability becomes an issue. The single most important reason for businesses to perish has been the goals deviating from the purpose. (Enron is just one glaring example)

The next important question is "How do we want to achieve what we want to achieve?"

At the fundamental level the response to this question will define the values that the business will cherish and live by. The values will have to align with vision and guide the manner in which the business would be done.

At the business life cycle level the question would help establish processes that will demonstrate the values the business will live by. The processes will be born out of the values and will have to be in line with the purpose of the business.

At a business execution level the processes should lead to strategies and at the same time they should be in line with the goals of the business.

The next critical question is the why of the vision which is the 'what' and values which is the 'how'. At a fundamental level the answers to the question would give clarity. Many a time leaders and those in positions of influence do not necessarily seem to be able to link why they are doing what they are doing with vision and values and this ambiguity tends to breed a certain level of disintegration. At a life cycle level the response would result in setting the standards the business would follow. The standards set should be born out of the clarity and be aligned with the processes established. At the execution level the response to the question would result in exploring various choices available which respects the standards and are in line with the strategies.

The critical question which is often dealt with is 'when to do what?' This at a fundamental level allows those in positions of influence to understand the context. When all the elements of the context are not understood the quality of decision suffers. An important element of the context is the clarity behind vision and values of the business. It becomes important to keep this in mind while evaluating each and every situation the business gets into.

At a business life cycle level the response to be question gives insight in to the need to be appropriate to the context and the standards set for the business. When either is taken for granted there will be a certain level of disintegration.

At the business execution level the response to the question would result in the final decision. The decision so taken should be aligned with the choices born out of strategies and will have to be appropriate to the context. When either of them is taken for granted then the mutation of the DNA of the decision begins which could result in various kinds of business diseases which can affect the business. A close look at the responses to each of the critical questions at different levels, gives one a deeper understanding of the evolution of the decisions that run complex business.

The matrix in Table 1 clearly indicates the need to think through several critical dimensions of a business while designing and executing the same. This analysis, while managing a business, helps those in positions of influence to remain appropriate in chaotic and complex situations. It gives a degree of stability in an extremely dynamic situation. The matrix emerges from the vision and moves on to finally finding the essence of it, through every decision taken during the course of the business. If any aspect of the journey between vision and decision is not taken into account or not thought through, then the business suffers in one way or the other. This in turn could result in its poor execution which would hurt the future of the business in the immediate future and the society in the long run.

Table 1
Decision Evolution Matrix

Questions	The fundamental level	Business life cycle level	Business execution level
What do we want to achieve? →	**Vision** →	**Purpose** →	**Goals**
How do we want to achieve? →	↓ **Values** →	↓ **Process** →	↓ **Strategies**
Why of what & how? →	↓ **Clarity** →	↓ **Standard** →	↓ **Choices**
When to do what? →	↓ **Context** →	↓ **Appropriateness** →	↓ **Decisions**

Source: © 2007 J.M. Sampath

Every decision taken during the course of the business should be retraceable to the vision of the business. Those in positions of influence should be able to see that every action of theirs contributes to achieving the ultimate vision, taking into consideration various elements that connect the decision to the vision. It is quite natural to experience a quality decision, taking all these elements intuitively into account. When a decision doesn't yield the desired results, it is almost certain that one or more elements would have been missed out or not considered enough. Thus it becomes important for leaders to be mindful of all these elements during decision-making. It has to become second nature for those making decisions to ensure everything done by everyone concerned with the business, is aligned with all the elements. The absence of this alignment results in conflicts that could affect the business and larger system in several ways (Table 2).

Over the years we have taken the values, process, clarity and standard for granted and all our goals, strategies & choices have been focused on short term benefits. We have manipulated every context and have been appropriate while making decisions only from the perspectives of short term gain and not from the perspective of long term sustainable growth. It is time that leaders, while making decisions, consider all the elements without having to compromise any and this would in the long run, help us reverse the degeneration that is already set in today's businesses.

Each of us need to understand that vision without values can be risky, values without vision will go nowhere and vision with values will result in the evolution of a world which will be more sustainable and easier to live in.

Table 2
Decision Traceability Matrix

	The fundamental elements	Project life cycle elements	Project execution elements
What do we want to achieve?	**Vision**	**Purpose**	**Goals**
How do we want to achieve?	**Values**	**Process**	**Strategies**
Why of what & how?	**Clarity**	**Standard**	**Choices**
When to do what?	**Context**	**Appropriateness**	**Decisions**

Source: © 2007 J.M. Sampath

PART 4

Debate: Republican Liberalism versus Market Liberalism

Peter Ulrich
Institute for Business Ethics
University of St. Gallen

Republican Liberalism versus Market Liberalism

Today's business life is full of tensions and conflicts between the "logic" of the market on the one hand and human, social and ecological demands on the other. In the name of "free market" and "free enterprise", the inherent necessities of the competitive economic system have been more or less released from moral inhibitions and institutional constraints in the last 200 years of modern development. The result of the ongoing political endeavor of "market liberalization" – on national level earlier, on the global level today – is a steadily rising productivity and economic growth. But the price we pay for this "economic progress" is also growing: what counts in the free market is what pays off for those investing their capital. This is the institutionalized *purpose* of the "system" of capitalism. All other (natural and human) resources have only the status of *means.* The "employment" of such means is calculated in terms of cost and has to be *minimized*, without regard for their intrinsic value, whereas the *return* on equity or investment has to be *maximized.*

Obviously, this "logic of the economic system" cannot be the whole answer to all our socio-economic problems, since economic activities are not an end in themselves but a means for the good life and coexistence of humans in society. For this reason the market economy has to be *civilized* in quite a literal sense: the constitutional task of a well-ordered society of free and equal citizens has always been to define the inviolable human and civic rights and to guarantee their primacy against all kinds of powerful infringements. *Free citizenship* is the core mission of every serious version of liberalism. This is something else and more than free markets. Thus, we have to make a fundamental difference between a *civilized* market *economy* and a total *market society*. In the latter, most social relations take the form of market relations so that civil society is reduced to market society (Polanyi 1991).

This is trivial in terms of political philosophy but not in terms of business ethics, as long as the latter is conceived of as an apolitical endeavor stressing primarily corporate ethics. The origins of the dominating concepts from the Anglo-American culture and their coincidence with the "given" and hardly questioned political framework of "free" market economy is probably not by pure chance.

The specific European contribution could be to reconsider business ethics as a piece of political philosophy and ethics. The challenge of *re-embedding* the "modern" market economy with its incredible productive powers into an equally modern society requires a systematic relationship among all the three disciplines involved: ethics, politics, and economics. This was the *Aristotelian triad* of practical philosophy. Classical political economy took it still for granted; but with the neoclassical paradigm the horizon was lost and "pure" and "autonomous" economics, detached from all ethical and political notions, was the consequence. Nevertheless, economics inevitably rests on explicit or implicit political presuppositions. The same is true for business ethics: the only choice concerning the political implications is either ignoring them (and this means: dealing with them in an ideological way) or reflecting on them with critical regard to their normative foundation. What a peculiar and symptomatic idea to separate the concept of free market economy from a well-ordered free and democratic society by trying directly to apply ethics to the business world, omitting the intermediate political level of the Aristotelian triad!

At first sight this seems to make only an academic difference. The contrary is true: it means a practical change of entrepreneurial self-understanding that really goes to the crucial points of today's socio-economic and politico-economic challenges. Business has to accept its basic need of orientation with respect to its societal function and legitimation before defining business integrity and corporate social responsibility. The usual "free enterprise" rhetoric for justifying conventional profit or shareholder-value maximization has then to be put away. Instead, business leaders will have to accept the primacy of societal values over business values and the resulting *co-responsibility* of business for the public order.

My concern is that we will not overcome the *mental barrier* to such a "civilized" understanding of entrepreneurship as long as the postulated primacy of political ethics over the logic of free markets is seen as a "leftist" worldview and not as what it really is: the core of true liberalism. What is needed is much more than the "the business case of business ethics" (i.e. a functional reduction of business ethics to a strategic factor of conventional business strategy). We need a comprehensive vista of the legitimate and reasonable function of business in and for a well-ordered society of free citizens, as it is the essential idea of *political liberalism* in the sense of John Rawls (1993).

Let me demonstrate the timely significance and the implications of the proposed embedding of business ethics in political-philosophical horizons in basic way. The economistic "*zeitgeist*" of our time was brought to the point in Bill Clinton's famous slogan: "It's the economy, stupid!" The politico-philosophical awareness I propose calls for a reformulation of that catch phrase: *It's not the economy, stupid, it's the society!*

The old economistic recipe supposes that most of our socio-economic problems, like mass unemployment and the growing disparity between rich and poor, find an economic solution: more "free" markets with harder competition. But these social problems are part of the *success story* of economic "liberalization" and "rationalization". For example: it is the inherent *purpose* of that program to set humans "free" from workplaces, i.e. to raise the labor productivity. Economic policies aiming at higher economic growth without changing the societal relations between all involved mean setting the fox to keep the geese. Our societal core problem is not a lack of productivity or economic growth. It is that we are not able to establish a reasonable societal order of free and equal citizens that is appropriate to an advanced economy. Instead of only fighting the symptoms of steadily rising productivity, especially by an ever "growing" social security system (hardly payable any longer), we should go back to basics and define the institutional preconditions of a well-ordered society in which free citizens are able to assert themselves with regard to all vital moments of a self-determined and good life, including its material base. Most people are powerless in socio-economic sense. They simply *lack real freedom*. The liberal order of an open society is in danger, if we do not stop the ongoing social disintegration. Why do we not finally accept this epochal challenge of the 21st century?

Of course, there are no chances to establish a better societal order without citizens – and business leaders! – who really want it, who understand it and who are able to acknowledge their resulting self-responsibility as well as their co-responsibility for the *res publica*. What matters is integrating this republican spirit within a liberal order towards the synthesis of a *republican liberalism* (Ulrich 2008).

Applied to business, this civic spirit can take the form of a *republican business ethos*. As economic agents the citizens have to integrate their "acquisitive intentions" (or "capitalist spirit" as explained by Max Weber 1930) in their civic sense and co-responsibility for a well-ordered "res publica" of really free citizens. As a result, they are interested in personal success or profit only as far as they – as citizens – can accept this as ethically and politically legitimate. Personal *integrity* means not to split off one's financial and business interests from one's civic identity. The contrast is obvious between this integrated republican-liberal identity and the "possessive individualism" (Macpherson 1962) on the line of Thomas Hobbes, which is nothing else than the concept of the person behind economic (market) liberalism. *Corporate citizenship* would accordingly mean much more than "corporate giving" after making the greatest possible profit with an unquestioned business model and more than "community involvement" as far apart as possible from the inherent normative content of this business model itself. Now, it means a basic attitude of republican co-responsibility of business for public concerns, especially for a "civilizing" political framework of the markets including all the implications for a company's own business policy.

To sum up, the fundamental difference between republican liberalism and market liberalism can be made by three basic points: first, as motivational base, an appropriate *civic virtue* with the consequence of a republican business ethos; second, as the constitutive order principle, fully established *civic rights* including (instead of excluding as until now) the socio-economic preconditions of a really free and decent living; and third, as an economic-ethical consequence, a *civilized market economy* (Ulrich 2008). Republican liberalism does not contradict an efficient market economy but has to be conceived as a societal precondition of legitimate and life-serving business activities.

References

Macpherson, C. B. (1962). *The Political Theory of Possessive Individualism: Hobbes to Locke.* Oxford, New York: Oxford University Press.

Polanyi, K. (1991). *The Great Transformation: The Political and Economic Origins of Our Time.* Boston: Beacon Press (orig. 1944).

Rawls, J. (1993). *Political Liberalism.* New York: Columbia University Press.

Ulrich, P. (2008). *Integrative Economic Ethics: Foundations of a Civilized Market Economy.* Cambridge, New York: Cambridge University Press. (In German: *Integrative Wirtschaftsethik: Grundlagen einer lebensdienlichen Ökonomie.* Haupt, Bern et al., 1st ed. 1997; 4th ed. 2008).

Weber, M. (1930). *The Protestant Ethic and the Spirit of Capitalism.* London: Unwin Hyman (in German orig. 1904).

Stephen B. Young
Global Executive Director
The Caux Round Table

Comments on Peter Ulrich's "Republican Liberalism versus Market Liberalism"

In his essay "Republican Liberalism versus Market Liberalism", Peter Ulrich synthesizes a number of important observations about social jurisprudence. I borrow the term "jurisprudence" from the law in order to emphasize the point of thinking about fundamental structures, the architecture of our civilization, the framing visions that lead to specific rules and concrete, behavioral norms in day-to-day life. Even more than law, society needs a just jurisprudence to give it authority and acceptable purpose.

After starting off with what I believe to be a false dichotomy between free markets and free citizenship, Ulrich makes several sound recommendations.

First, he calls for integrating Aristotelian politics with ethics and the economy. Ethics and politics are really little more than different sections of the same continuum of human concern. Both attempt to link the self with the other: ethics by giving us internal constraints on how we use our powers vis-à-vis others, and politics by structuring our communal power arrangements. In Aristotelian terms, individual ethics arise from an "ethos" which is social and cultural in origins and continuity and politics creates and sustains the "ethos" of the community.

So simplistically, bad politics leads to bad ethos, which in turn brings on bad ethics. For example, in a society filled with violence and mistrust, it is hard to get individuals to act self-lessly.

But there is another, I think, deeper implication to Ulrich's point on the need for politics. Capitalism – of whatever variety – only comes to fruition within the context of a given set of social capital accumulations. Endemic poverty in some societies is a direct function of the social capital accumulation of such societies. It pays us great intellectual dividends to remember that modern, self-sustaining capitalism, per se, was born in cultural conditions of Calvinism in the limited social surroundings of The Netherlands, Scotland and England in the late 16th and 17th centuries. The analytical power than comes from this remembrance of times past is a conviction that business succeeds in certain cultural conditions and not in others. Politics, it would then seem, must play an important role in creating such conditions.

So, as Ulrich argues, a totally Hobbesian world combined with a purist, devil-take-the-hindmost, adversarial free market set of economic rules will not produce a robust capitalism.

I would only add as a point for further reflection that 18th century Whigs, among whom we should certainly count Adam Smith, would only agree with Ulrich that bad politics leads to bad economics.

Second, Ulrich calls upon business to accept its basic need of an orientation with respect to its social function. With this I entirely agree as well. The modern business corporation is a social creation, structured by complex contracts reflecting mutual agreements and supported by laws. Business exists to serve; its function is to create wealth for society. It has an office of servitude to perform, or to be more respectful, an office of stewardship.

With Cicero's work on ethics De Officiis, we see in his use of old Latin that the position of holding an office – of having duties – encumbers an individual self (or individual firm for that matter) with obligations and responsibilities to others. The morality of holding office is assuming duties of care and loyalty for some set of beneficiaries who will be affected by the power given to the position.

A business firm has economic, social and political power. Its actions are not without consequence to others. It spawns so-called public "goods" and "bads". And so, not inappropriately in my judgment, it is called upon as Ulrich does to internalize a broader range of goals and objectives than is contemplated by short-term maximization of cash income.

Business is a means; profit is not an end but only a measuring rod of successful stewardship. Profits demonstrate that the business is well-run and is fulfilling its function of wealth-creation. Running a business at a loss demonstrates to the contrary that the firm is destroying wealth by taking capital funds from others in the form of a subsidy of its activities.

But, Ulrich I think makes a category mistake when he conflates the business enterprise with those who have responsibility for a public order. Business has a limited set of responsibilities if it is to be true to its office. Business is not government; it should not be run on principles of quid-pro-quo, muddling-through, or grid-lock until a consensus emerges among the citizenry as to what to do. The point of a business is to take risks – entrepreneurial risks – and doing so mandates making decisions and cutting opportunity costs. In responding to the desires of some, business must reject the desires of others.

Government has its own office to perform, separate from that assigned to business. Government needs its own set of ethical principles to keep its use of power just and responsible.

And, though Ulrich did not mention it, I would add that civil society has its independent role to play in maximizing the gains to be had from modern civilization.

But when we are assessing the advantages and disadvantages of our versions of civilization, we should not forget the golden eggs when we critique the goose. Capitalism, free markets, business enterprise, have given us health, wealth and power that our ancestors never imagined. Bringing to market cheap cotton cloth, flush toilets, domestic water lines and waste sewers in the early 19th century saved tens of thousands of lives, especially those of small children. Today in advanced economies few are needed to produce vast quantities of food that can be sold to the masses. Well-designed clothes sold for modest prices have obliterated most public and superficial class distinctions that traditional societies fought over. And, the implications for humanity of each one of us being able to live in a "global village" due to the low cost of digital communications are quite revolutionary. The contributions of the fruits of capitalism for social justice are many and should not be overlooked in the first instance.

The critique of our civilization, I would argue, should be pointed more at things of the mind and the spirit than at forces and means of capitalist production. What Max Weber called "the disenchantment of the world" as a consequence of a European Enlightement fixation with a certain kind of disembodied rationalism may have more to do with our current social and cultural discontents than free market practices.

We are still captains of our souls, if not masters of our fates, and, as free citizens of the world, we can choose freely among religions and ideologies and even mysticisms for a sense of purpose and service in life.

The republican liberalism advocated by Ulrich sounds to me very much like the classical Whig social philosophy which gave us Adam Smith in economics, Blackstone in law (the "Rule of Law") and constitutionalism in England and the United States. It was individualistic but also communal in the tradition of Cicero where individuals were expected to live up to the obligations inherent in their station, be that station high or low.

What Ulrich refers to as "Market Liberalism", I would argue is not a liberalism at all for therein the individual has been conceptually freed from responsibilities and duties and confirmed only in his or her Hobbesian selfishness. This is Social Darwinism, or the philosophy of unredeemed libertarianism, a dark and sad vision of human life as "red in tooth and claw."

Jean-Pierre Galavielle
Université Paris1
Centre d'Economie de la Sorbonne,
Paris

Who Can Civilize the Market?

The question is important if we accept to place it within the framework of recent history that has given birth, on the occasion of the crisis of Fordism, to the idea of *ungovernability,* implying that public authorities were no longer capable of bearing their responsibilities.

Here we meet with the famous question "*How to govern without government ?*" However, somebody will have to take care! Why not enterprise, NGOs or charitable institutions, in other words "Civil Society". The private sector and civil society are entrusted with tasks that belonged to public power. Here emerges the particularly ambiguous theme of civil society which, after several centuries of semantic history, finally stands to-day for the ultra-liberal slogan "anything but the State", and henceforth "appears as the new' must' of contemporaneous discourse" (Lochak 1986). We may note that this ultra-liberal orientation adopts the old Marxist thesis of the decline of the State!

Even if this answer may appear seductive, it nevertheless carries with it the fundamental question of the "democratic legitimacy of non-elected authorities receiving a mandate through public delegation" (CAE 2002). These organizations justify their legitimacy via the auto-proclamation of their expertise. The expert, *initially an instrument of power, becomes power himself* to the detriment of the democratically elected politician.

The questioning of the traditional model that attributes to political authorities the management of public matters, stands for a crisis of ungovernability. It is rooted in the hypothesis of the incapacity of the State to direct public matters, and has to be characterized as follows:

- Public authorities no longer hold the monopoly of public action and of the responsibility that is attached to it.
- Private actors (associations, coordinated bodies, communitarian pressure groups, etc.), claim to be associated with regard to decision processes.
- The responsibility of the State is progressively transferred towards civil society and the market, while the frontiers between the public sector and the private sector are being displaced.
- There is a growing gap between the vision of political apparatuses and the life of citizens.

This loss of legitimacy of public power is certainly a crisis according to J. Habermas (1978) who, proceeding via an analogy with medicine, has defined it as "the decisive phase of an illness that permits to know whether the strength of the organism itself will suffice to bring back health… For social sciences, the crisis emerges when the structure of a social system… admits less possibilities of solution than the system needs in order to maintain itself…The crisis phenomena owe their objectiveness to the fact that they are born out of regulatory problems that have not found a solution" (Habermas 1978).

In other words, the crisis of governability appears due to a loss of legitimacy of the socio-political system. Thus, several forms of crises may occur: crisis of rationality; crisis of legitimation (loss of loyalty of certain actors); socio-cultural crisis (loss of traditions, of ideological anchorage). The conjunction of these forms of crises leads to the creation of a sort of "Brownian motion" (Habermas 1978)).

If this is so, does the fatality of ungovernability inexorably lead to the dictatorship of the "Government of things"?

1. Evaluation and Policy

For some years, evaluation has installed itself in all possible fields. There is no longer any activity that would not be concerned by it, so that it has become a "way of functioning of the State". For some, it might even establish a demarcation line between "democracies founded on law and democracies founded on contract" The reason for this is simple: "the first permit revolt against the law, whereas the second have had the consent of the citizen with regard to the process, so that he cannot resort to revolt" (Ruby).

Furthermore, this frenzy of measurement constitutes an encouragement to repose all appreciations and/or decisions on the principle of the law of the strongest. The diffusion of these evaluations and the formulation of value judgements do no longer refer to the intrinsic quality of a presentation but to its commercial success in terms of market share or sales volume.

Thus, there is the imperceptible diffusion of the dominance of "generalized (evaluation that) places its hand on everything" and becomes "supreme expertise". The evaluator thus becomes "the mouthpiece of things" (Milner 2005). However, when the managers report on their action, or try to justify it, they affirm that they often find themselves in an awkward predicament because of the obligation to achieve results on the one hand, and the law of the market on the other.

2. Accounting Standards and Results

Results are evaluated by statistical and/or accounting instruments, the scientific character of which is proclaimed by a consensus of experts, without being actuality demonstrated. These standards that are presented by experts in this field, are all the more reassuring as the poster of their scientific nature induces the reliability and the trustworthiness of the announcements they transmit (Marchesnay 1998). They inspire confidence.

Accountancy has become the main instrument of financial communication for enterprise within the framework of globalization of financial markets. This procedure led to the creation of an abyss between productive and financial reality and the reality of the accounts that are supposed to translate, because of the extraordinary variety of rules for registration:

- are the costs incurred on the occasion of the constitution of a client portfolio, investments or operating expenses, when this portfolio may become, after a certain time, a negotiable asset?;
- the evaluation of 'faire value' and off-balance sheet commitments; 'Special purpose entities' that are not bound by consolidation rules; anticipation or carry-over of sales; *management of provisions and reserves* ("cookie jars") *in order to facilitate the regulation of future results;*
- the opaqueness of documents, audits and comments accompanying accounting documents;
- the invention of indicators with scientific names that don't mean much: ROE[1] EBITDA[2], etc;
- the absolute priority granted to the orders of the shareholder.

The FASB[3] admits that "the standards permit accounting engineering in order to structure transactions by avoiding the rules and by circumscribing the spirit of the standards". *Harvey Pitt*[4] *draws himself the following conclusion:* "The development of accountancy based on rules has led to the utilization of financial techniques defined only for the purpose of realizing certain accounting objectives".

More generally, "under the cover of a technique … accounting is in reality an entity for social constructions historically dated and implying economic effects… If accountancy has a history linked to economic and social conditions, this means that it cannot pretend to offer scientific products, that it may not be trapped into a limitless atemporariness and that it cannot present a universal character" (Capron 2005).

3. The Myth of the "Market Law"

Another component of the government of things is the "Market Law". On first sight, one may consider it as the representation of the confrontation between supply, demand and the prices that thus evolve. It owes its scientific character to its capacity to verify, explain ant anticipate situations or phenomena in accordance with entities of hypotheses formulated previously. On this basis, it claims universality. Nevertheless, one may ask what is the sense of the expression: "This is market law"!

The debate concerning the question of the theorization of financial markets is very significant. Successively, we move on from a Gaussian approach (Quételet, Régnault[5], Bachelier[6]) to a vision of Brownian motion and from there on to chaos.

J. M. Keynes (1921) quite frequently quotes recourse to the Normal Law as the fruit of a convention of normality he calls himself a "financial convention", meaning that all market participants accept the legitimacy of the price indicated (Keynes 1921).

In 1959, the astrophysician M. F. M. Osborne publishes an article entitled "Brownian Motion in the Stock Market". He demonstrates that statistical analysis applied to the investigation of an entity of molecules is perfectly valid for stock exchange quotations! Later on, Benoît Mandelbrot (1973) observes that these stock exchange quotations fluctuate in such an erratic way that they do not obey any identifiable statistical law like the Normal Law (Mandelbrot 1973). This is the very basis of the famous theory of fractals, close to chaos theory.

Reference to market law leaves many questions unanswered:

- What is the scientific character of a convention?
- Is an efficient market necessarily a rational market (Aglietta 1995)?
- Is the information it conveys always pertinent and representative of the real economy?
- By virtue of what, the suggested rationality of market participants will lead to globally Brownian results?
- How to explain price anomalies?

The scientific character of market Law is most doubtful and its universality has still to be demonstrated. It is difficult not to think in this connection of the "Invisible Hand" about which J. Robinson wrote that it might "work by strangulation" (Robinson 1946).

There remains the law of the strongest. This perspective which is much more realistic for an economy largely dominated by supply, recalls certain realities: natural selection, the choice of short-term satisfaction for the shareholder, risk aversion, dilution of responsibility, etc.

Everything that has been said so far intends to demonstrate the calculated deceit hidden behind the "Government of things". Reference to its dictatorship is based on self-realizing anticipation. It is not possible to affirm the supposed virtues of a way of functioning of the economy, largely develop its instruments and consider at the same time their pertinence as a fatality. *We are nowhere near objects given to us or forced upon us by nature, such as the Sun or universal gravitation.*

4. Which Legitimacy?

There remains the question of the legitimacy enjoyed by independent or private organizations, that makes them pretend to regulate society.

The emergence and the dissemination of the theme of responsibility are the consequence of the loss of confidence of societies in institutions habitually in charge of public regulation, whereas the development of new technologies accredits the idea of zero risk. This defiance with regard to public institutions forms the culture-medium that gives rise very rapidly to other pretenders to societal regulation: NGO, Enterprise, charitable Associations of any kind, New social-economic movements, communitarianism, etc. All these organizations are voluntarily kept apart from any external legitimate political control. These substitutes are the fruits of an ideology of the inefficacity of the State, erect expertise into a decision model, are grounded in evaluations, produce standards of all kinds, and when they lack arguments, refer to the law of the market.

From here, it is easy to understand that this affirmation of expertise may at the beginning carry the illusion of legitimacy, whereas it does not really pass the test of the respect of democracy. Societal responsibility constructs itself to-day within a closed field where the expert and the politician are facing each other. One must also underline the fact that the politician has to assume a large share of responsibility, because to start with he himself made the choices that formed the basis of his own renunciation. Is it thus not amazing that he has become, due to unforeseeable reversals of political majorities, the object of the contempt of institutions that he wanted himself to be independent and in a certain way uncontrollable?

How may one claim a responsibility, considering that one deprives oneself of the means to exercise it?

The same spirit prevails when on the one hand public expenses and tax revenues are being restrained and deficits enlarged in order to more easily proclaim their unsustainability, and when on the other hand there are organized enormous campaigns of charitable collections in order to finance medical research or to help unadapted childhood or indigence.

Indeed, *the problem of the imbrication of the market and democracy within contemporary capitalism* (Aglietta and Rebérioux 2004) will not be resolved as long as there will remain the myth of the dictatorship of the Government of things. "Civilize the market" means that the politician must urgently regain his place as a democratic regulator of production, exchange and distribution of wealth.

Notes

1 Return on Equity.

2 Earnings Befor Interest Taxes Depreciation Amortization.

3 Financial Accounting Standard Board.

4 Harvey L. Pitt: President of the United States Securities and Exchange Commission.(SEC)

5 J. Regnault: « les variations de la bourse sont soumises à des lois mathématiques immuables ! » (trad. « stock exchange variations are subject to unmovable mathematical laws ! »)

6 L. Bachelier "the mathematical expectation of the speculator is zero".

References

Aglietta, M. (1995). *Macroéconomie financière.* Coll. Repères, La Découverte.

Aglietta, M. & Reberioux, A. (2004). *Dérives du capitalisme financier.* Albin Michel Économie.

C.A.E. (2002), P. Jacquet, J. Pisani-Ferry, L. Tubiana. Gouvernance Mondiale, Rapport du Conseil d'Analyse Économique. *La documentation Française.* May 2002.

Capron, M. et al. (2005). *Les normes comptables internationales, instruments du capitalisme financier.* Coll. Entreprises et Société. La découverte.

Chevalier, J. et al. (1986). *La Société Civile.* P.U.F 1986. Rangeon. F., "Société Civile: Histoire d'un mot" (pp. 9–32); Lochak, D., "La société civile du concept au gadget" (pp. 44–75); Drai, R., "Préalable éthique de la société civile, disjonctions et conjonctions" (pp. 76–106).

Habermas, J. (1978). *Raison et légitimité.* Payot.

Keynes, J. M. (1921). *A teatise on probability.* London: Mac Millan.

Magnette, P. (2000). "Quis custodes custodiet ? La banque centrale Européenne entre indépendance et responsabilité". *Cahiers Européens de Sciences Po,* 1/2000.

Mandelbrot, B, (1973).
(1) "Formes nouvelles du hasard dans les sciences",
(2) "Le syndrome de la variance infinie et ses rapports avec la discontinuité des prix",
(3) "Le problème de la réalité des cycles lents et le syndrome de joseph". *Économie*
Appliquée. Archives de l'ISEA, Tome XXVI 1973. Librairie Droz Genève.

Mandelbrot, B. (1995). *Les objets fractals.* Flamarion, Coll. Champs.

Mandelbrot, B. (1997). *Fractales, Hasard et Finances.* Flamarion, Coll. Champs.

Marchesnay, M. (1998). Confiances et logiques entrepreneuriales. *Économie et Sociétés,*

Sciences de Gestion, Série S.G. No. 8–9/1998, pp. 99–117.

Milner, J-C. (2005). *La politique des choses.* Navarin.

Robinson, J. (1946). The Pure Theory of International Trade. *Review of Economic Studies,* Vol. 14, No. 2. (1946–1947), pp. 98–1 12.

Ruby, C. (2004). "Le paradigme de l'évaluation", *EspacesTemps.net,* 28/10/2004. Note de lecture à propos de "Voulez-vous être évalué ?" J-A Miller et J-C Milner, Grasset. Paris, coll. Figures, www.espacestemps.net/document735.html

Alpar Losoncz
University of Novy Sad

Is Ethics Integral?

P. Ulrich makes a lot of insightful suggestions considering the status of business ethics. In addition, his line of reasoning is linked to an ambitious program of republican business challenging the mainstream thinking. In short, we are confronted with a tightly reasoned argumentation that combines the critical diagnosis and advocating of the reconceptualizing of business ethics.

The critical analysis refers to the misdirected, perverted relationships between the ends and the means under the recent market liberalization that leads to the reduced and ideological understanding of the market. Evidently, in appeal to efficiency we are witnessing the expansion of the market or the strengthening of the market-dependence in finance-driven capitalism. Yet attempting to come to grips with this sort of tendency need not reify it.

There are many considerations within the contemporary thinking that put emphasis on the different meanings connected to the functioning of market. Thus, there is essential distinction between the market as an opportunity-driven economy of efficiency responding to periodic ups and downs of various economic factors etc. and the totalizing market imperatives. The promotion of the market as self-reinforcing loops of business activity, microeconomic driver, and institutionally constrained business game is at variance with the interpreting of the market as a self-referential normative domain. The market economy is not to be equated with the market society, that is to say, with the idea that market provides normative horizons for the socio-economic fields. Actually, the theory of market must be "historically specific" as it shows how specific set of institutions, characteristic of market economy, such as profit imperative, depends on the larger systematic environment in which an economy is embedded – its boundary conditions. These include political and economic dimensions as well as environmental and resource/energy constraints.

Ulrich especially addresses the issue of business ethics: does business ethics as a new form of thought contribute to the recent trends of depolitization with its "apolitical endevour"? Does it miss the far-reaching critical attitude toward the "failure of the modern experiments" (Schumacher) reflected in the current trends, too? Is it too permissive in relation to the instrumentalization of its own perspective as a factor of profit making? Is the business ethics characterized by an overadaptation to the neoliberal agenda of "strong business rights-weak responsibilities"? Ulrich traces us back to the roots of liberalism and calls attention to the substantial alteration concerning the "liberalism"; as a matter of fact, he envisages the shift from the "possessive" to the "republican liberalism.

It will not be doubted that this operation would require a stronger sense of community than now prevails. The republicanism articulates the continuity between the private and public ethics; in fact, it questions the dividing of the freedom, and requires a strong sense of common good. The indivisibility of the freedom is in line with the claiming of the effective and participative citizenship. But how to reach the phase of the republican business? For it seems likely that if the business can reach it, or reach any substantial installment of it, its way along the road to reaching it will have made it capable of running it, or at least less incapable than the business now is. For example individual property rights can civilize self-interest of business, but they need not do so. In fact, in many situations the property rights can only be effective to the extent that they are not themselves the expression of self-interest. Because self-interest invariably seeks to capture the forces which can discipline it, the result ended with power concentrated in relatively few hands, not responsible to anyone or to any criterion other than its own preservation and expansion. How can we reach the situation of the republican property regimes, for instance?

It seems to me that we could accept the importance of inherent contradiction of the "historically liberalism" accentuated by C. B. Macpherson.[1] Therefore, we should treat the inherent ambiguity between the historical forms of market freedom ("Market man") with the self-developmental freedom. The first mark of liberalism brings to the fore the idea that the traditional focus of, especially non-mainstream, political economy on power is still of paramount importance. The second aspect in my view goes back to much older practice of the "relation to self" which calls for the analysis of ethical self-constitution. In this sense, business ethics could be linked to the possible way of the "conduct oneself".

Ulrich suggests that the business ethics should be more active in the reformulation of citizenship. This is clearly related to the idea that there is a need to make stronger the ties between the business ethics and political philosophy. In reality, there are attempts in business ethics to shed light on the mentioned relationship[2].

It is clear that the republican community *qua* politico-business community must be concerned with satisfying more than common economic needs. I do not deny the importance of the business ethics in the context of the political sphere, but, from my perspective the business ethics could only *indirectly* be connected to the articulation of the political sphere. The argumentation of the business ethics could not be directly translated into the political field. The business domain is only one order among others in a given social formation. Thus, it is just part of society and only indirectly is charged with responsibility for maintaining the cohesion of the wider society. In the society whose common interest is constituted, reproduced and transformed in and through heterogeneous processes we should take into account the autonomy of different spheres such as politics and economy. For this reason, I am skeptical in relation to any concept of "corporative citizenship", in spite of the fact that there is no complete separation between the political and economic domains. After all, we are forced to permanently explore and evaluate the changing relationships, moreover, the non-correspondences amongst the economic, non-economic forces, and ethical considerations.

Note

1 Macpherson, C. B. (1977). *The Life and Times of Liberal Democracy.* Oxford, New York: Oxford University Press, 1. See the articulation of the market power: W. Kingston, A spectre is haunting the world – the spectre of global capitalism, *J Evol Econ*, 2000, 10: 83–108.

2 For example, Kaler, J. (2000). Positioning Business Ethics in Relation to Management and Political Philosophy. *Journal of Business Ethics*, 24: 257–272.

Gerhold K. Becker
Assumption University Bangkok and
Hong Kong Baptist University

Which Role for Business Ethics? Some Reflections on Peter Ulrich's Statement

Ulrich's overly brief programmatic summary of his conception of an integrated economic ethics is the result of his debate over the last twenty years about what he regards as the two dominant models of business ethics: moral economics and applied (business) ethics. In Ulrich's reading, the former locates ethics within the strategic framework of business practice, while the latter merely attempts to remedy moral shortcomings in actual business conduct. Both seem uncritically to accept the parameters of free market economy along the lines of orthodox economic theory. Therefore they both fail by ignoring the need to establish the primacy of ethics in business on the basis of a "comprehensive vista" that restores politics as an integrative constituent in the interplay between economics and ethics and grounds questions of the good in a firm conception of right. Ulrich paints the picture of the alternative "we" are facing in stark strokes of black and white: Either the market can be "civilized" or civil society will be reduced to "a total market society." Such civilizing of the market requires, above all, a consensus among all stakeholders (and certainly not only among shareholders) on "the legitimate and reasonable function of business in and for a well-ordered society of free citizens."

Up to now Ulrich focused his critique mainly on the rival model of economic ethics and its corresponding economic theory. With the imminent publication in English of his magnum opus *Integrative Wirtschaftsethik* (1st ed. 1997, 3rd ed. 2001) he issues a wake-up call to applied business ethicists, presumably of Anglo-American provenance, who he fears are (still) largely unaware of the grave danger they are facing and in need of his helping hand to put up last-minute resistance. He appears in the role of a knight in shining armor who single-handedly brings European troops (and their "specific European contribution") to the battle field to relieve apparently inept defenders under attack by the unfettered forces of the market and its inhumane logic.

While the overall goals of Ulrich's ethics are persuasive up to a point, its test case as a highly ambitious discourse-based universal ethics is its practical feasibility within the business context of cultural diversity and globalization. After all, business ethics of any type would miss its target if it were to substitute comprehensive visions of substantive ethics for ethical guidance to business practice and the continuous evolution of regulatory background institutions at domestic and international levels. As regional ethics with a specific subject matter, business ethics shares the precarious conditions of related ethics disciplines that would be blind without substantive conception of the good but ineffective without mechanisms of application.

Limit of space allows neither to do justice to the complexity and sophistication of Ulrich's proposal for an integrative economic ethics (particularly as it has been presented in his major publications), nor to sufficiently expound the critical questions it provokes. The following remarks can only point at some issues deserving further investigation.

Ulrich employs the somewhat ambiguous strategy of first claiming the existence of a dichotomy he then denies: amoral business and economics on the one side, ethics, virtues, and moral norms on the other, societal, side. Similarly, he sees most business ethicists accepting this division (at least indirectly) by merely promoting a niche for ethics in business activities as a further strategic element for profit maximization. He launches his critique of such division from two angles: an anticipated comprehensive conception of a well-ordered society and the analysis of the factual interdependence between economics and society. Despite all rhetoric against amoral business his analysis suggests that business in fact never lacks moral background institutions and embedded normative standards, and would be impossible without them. The amorality of business may therefore indeed be a myth that is still characteristic of certain types of economic theory but not of business culture in general, despite all the scandals and widespread moral shortcomings of individuals. His critique therefore amounts to not much more than a reminder to all concerned that the business of business is not business but human flourishing, and that this cannot be had against morality.

Ulrich's moral rehabilitation of business as such (in contrast to individuals in business) does not, however, extend to applied business ethics, and this is no oversight. His ambitious design of an integrated economic ethics is based on the comprehensive vision of a well-ordered society, and in his view it is exactly the significance of such a vision that is not accorded a proper role in conventional business ethics, particularly in its applied version. Yet it is doubtful whether this charge can be substantiated, although it is certainly correct to assume that ethicists disagree on Ulrich's conception of a well-ordered society both in its substantive content and functional role for business ethics.

Ulrich's somewhat impoverished understanding of applied ethics as mere problem-fixing lets him underestimate conceptual and practical problems of his own theory. One such problem concerns its role and placement within the ever-growing field of ethical sub-disciplines and their presupposed respective concepts of the good life. While, for example, the ethical discourses on biotechnology, medicine, health-care, or the implications of human activities for nature and environment clearly overlap with concerns in business ethics, it is unclear what role

Ulrich can assign to them in his theory. On various occasions he seems to present his conception of an integrated economic ethics as the super-theory that can absorb (or "integrate") all rivals and transform itself into a foundational theory of the good society.

A case in point is the underdetermined relationship between the various types of rights (basic rights, civic rights, rights people are supposed to have as economic citizens, and human rights), their moral foundation, their hierarchical ranking within Ulrich's conception of the good life, as well as their institutional back-up in positive law. It is even less obvious which institutionalized procedures across nations will be required for their endorsement and their practical implementation in the global world of business. In other words, the presupposed subjects ("we") entrusted with bringing about the civilized market economy Ulrich envisages remain somewhat obscure. In the "we" he apparently lumps together government officials (negotiating trade agreements), business executives, economic citizens, and other stakeholders (including business ethicists) whose foremost moral duty it is to embrace "an appropriate civic virtue" so that "we" will be able "to establish a reasonable societal order of free and equal citizens." Unless Ulrich specifies the exact roles of all those responsible for bringing the moral point of view to bear on the deliberative processes of defining the social function of business (domestically and globally) his theory will amount to little more than another appeal for civic and moral virtues in the conduct of business.

Closely related are concerns about the heuristic and methodological value of utopian projections and visions in business ethics. One will recall that conceptions of the well-ordered society inspired ethical thought throughout human history and were the driving forces in the process of civilization without ever achieving consensus among "all stakeholders." In the age of globalization, the ever growing circle of the "affected" is no longer restricted to nation states, their citizens, and domestic policies, but includes all people, regardless of social status, communicative ability or economic power; – it may even transcend humanity and include "nature" (ecological systems, higher animals). The persuasiveness and cognitive appeal associated with political conceptions from Plato (Politeia, Laws), to Kant, Marx, and Rawls, has up to now not translated into universal agreement or "civilized" economic systems purely through deliberative human action. On the contrary, market developments appear largely, if not exclusively, determined by a conglomerate of forces of chance, luck, and self-interest (both at individual, national, and international levels) that only Hegel's optimistic reading could identify with hidden operations of reason (List der Vernunft). In a divisive world of competing political systems and their underlying visions of the good, which usually conflict anyway with political reality, the chances for just and fair international agreements on finance, commerce, and trade are as slim as ever.

If this suggests a more modest (or maybe pessimistic) attitude towards the capability of business ethics, it may nevertheless facilitate a more accurate picture of versions of applied ethics. While they deliberately refrain from projecting large vistas of comprehensive conceptions of the well-ordered society, they are certainly rooted in strong moral convictions that enable them to identify wrong developments and propose feasible alternatives. Ulrich, too, recognizes the need for constructive involvement of ethics in daily business practice before the agreed-upon vision of the good life has taken shape. He says, however, surprisingly little about the normative foundation of such intervention and even less about the practicalities of uniting all affected in the realization of the moral vision. Instead he seems optimistically to rely on the cognitive force of the good argument in the sterilized thought laboratory of discourse ethics that allows access only to morally competent citizens stripped of all cultural characteristics. The problems this construct raises, particularly in view of the ineliminable pluralism of conceptions of the good in a global world and their grounding of conceptions of right, are too complex but also too well known to be discussed within the confines of this comment.

In conclusion, Ulrich's proposal for the readjustment of the multifarious discourse in business ethics towards a reflection of its fundamentals is timely and welcome, particularly as it raises a number of questions that will, hopefully, stir up a more focused debate about the realistic, long-term goals of business ethics and their socio-political, cultural, and philosophical presuppositions.

Peter Ulrich
Institute for Business Ethics
University of St. Gallen

Reply: Republican Liberalism and Its Implications for Business Ethics

Authors of short programmatic papers on broad subject-matters usually incur a feeling of having said almost nothing and omitted so many points that should have been explained. All the more it is a pleasant experience if dialogue partners nevertheless grasp almost everything. Such an astonishing understanding probably rests on basically common thought patterns. And this seems to be the case at least with three of my appreciated commentators (J.-P. Galavielle, A. Losoncz, and S.B. Young), but not so much with G.K. Becker whose perspective differs considerably from mine. Hence it is understandable that he complains of an "overly brief programmatic summary" of my conception and partly reacts with trivializing metaphors and some incorrect interpretations. There will not be enough room for clarifying all of them but at least – and with the help of the three other comments – I try to explicate some of the systematic implications of republican liberalism for business ethics.

As far as I see, the conceptual ties between Losoncz, Galavielle, Young and me rest on the guiding idea of the normative (but, of course, not factual) primacy of the principles of a well-ordered society of free and equal citizens over the factual (and at the same time implicitly normative) logic of "free" markets. These idea implies a specific relationship between the conceived responsibility of business and the "public use of reason" (Kant 1991, Rawls 1993: 212ff.) in civil society, which is understood here in its comprehensive meaning as the *societas civilis*, inclusive of the rule of law and the public order, not apart from it (Kaldor 2004: 7ff.). The politico-philosophical debate on these issues has reached a remarkable level during the last decades. There is a lot to learn from the results; it is time that business ethics takes notice of that in a more serious way than until now. That's what I try, and I cannot find any argument by Becker who merely asserts that "it is certainly correct to assume that ethicists (in general, P.U.?) disagree on Ulrich's conception of a well-ordered society". Of course, I make a systematic difference between Rawlsian political liberalism and my own conception of republican liberalism (see Ulrich 2008, Sect. 8.1). Dagger (1997) has developed a concept of republican liberalism very similar to mine, at the same time and without mutual knowledge, but he is not concerned with the socio-economic side of the problem. And this is exactly the specific challenge and task for integrative economic ethics!

The problem with Rawls' conception is that it lacks a clear and sustainable delimitation from market liberalism because he does not consider the structural partiality of free markets, though this partiality is incompatible with his own postulate that the basic political order of a liberal society has to be neutral with respect to different concepts of a good life (*cf.* Ulrich 2008, Sect. 7.2). This is a serious objection of a specific economic-ethical kind. It marks the actual core problem of liberal societies all the more as there is a real tendency of misusing the term 'liberalism' for the purpose of euphemistic masking of an economistic ideology of free market capitalism, at least in the German-speaking countries of continental Europe. And, of course, the latter doctrine is not a kind of true liberalism at all in a philosophical sense of the term, as Stephen Young has noticed at the end of his comment.

Now, the political order of a truly liberal society does neither fall from heaven nor result from mutual exchange of private advantage between *homines oeconomici*. Without a certain degree of citizen's morality, we cannot expect a fair and just public order as a result of democratic procedures. Garbage in, garbage out. Under an ideal horizon of orientation, the necessary civic ethos should be a post-conventional one in the meaning of Kohlberg's (1981) development stages of moral consciousness; otherwise we might fall back in a helpless clash of cultural value-traditions and life scripts. Republican liberalism is the proposal of a balanced conception that includes both a strong but neutral political order enabling a "reasonable pluralism" (Rawls 1993: 24) of conceptions of the good, and responsible citizens with a *republican ethos*, which means in a post-conventional sense: free citizens who recognize their shared responsibility for the well-ordered *res publica* and do not cleave off their business life from their civic identity and integrity. The resulting idea is *economic citizenship*, including an individual moment of citizenship responsibilities and an institutional complement of strong citizenship rights along Marshall's (1987) path of development (see sect. 7.3 in Ulrich 2008). This integrative concept of a modern civic business ethos for persons and companies ("corporate citizenship" in a literal sense), supported by a conducive institutional backing, is exactly the bridge over the categorical dichotomy between free markets and free citizenship as disputed by Young and Becker.

By the way, Becker confuses my analytical differentiation between pure economic rationality (i.e. the normative logic of the market) and ethical reason (i.e. the normative logic of interpersonal relations between free citizens) with the two-world concept of "amoral business and economics on the one side, ethics, virtues, and moral norms on the other, societal, side", which is not mine. This "myth of amoral business" (DeGeorge 1990: 3ff.) and the corresponding scientistic self-misunderstanding of "pure economics" as value-free and interest-neutral (i.e. as completely detached from ethics and politics) are just the objects of my criticism. Integrative economic ethics makes the point that normativity is always and unavoidably inherent *within* and at the bottom of "pure" economic reasoning: the latter approach implies a certain political doctrine based on a mysterious implicit ethic called "invisible hand" (allegedly making personal morality dispensable) that has to be brought into the light of rational ethical and political argumentation and justification. This is the politico-philosophical core of a civilized market economy open enough for deliberative and democratic political design in many variants but immune against any economistic reduction to a total market society (which, of course, is only an ideal-typical, life-practically absurd border case).

To demystify the *zeitgeist* syndrome of normatively deep-rooted economism is surely the most ambitious but the decisive challenge of economic and business ethics as far as it wants to be a systematic part of the solution and not a symptom of the problem. Therefore, neither the defence of the false "morality of the market" (moral economics) nor an approach of business ethics as applied ethics is adequate. The core problem of the latter is that it is conceived as a tool of social engineering for ethically good ends under the "given" business conditions of the existing market economy without reflecting these conditions themselves to the end. Ethics *within* the frame of free market capitalism is no longer enough today – what is needed is a comprehensive ethics *of* the preconditions of a legitimate and life-serving market economy. Becker may call that "utopian" and not useful for the "practicalities" of real business, and he is right so far, although I do not see a contradiction to work as a business ethicist on both levels, that of comprehensive vistas as orientation for a good politico-economic development and, at the same time, on the pragmatic level of business integrity and responsibility (see e.g. Maak and Ulrich 2007). But to do the latter without the former is in danger of losing applied ethics from the indispensable regulative ideas that give us well-founded normative orientation. There is indeed a kind of "epistemological interdependence" (Kaler 2000: 266ff.) of responsible business activities and their legitimating institutional framework. What in addition to this republican liberalism shows forth and stresses is the *republican co-responsibility* of "private" business for this institutional framework since, as already said, good regulatory policies do not fall from heaven. (This does not mean "conflating" business and government, as Young worries; however, we know from "real politics" that the regulatory framework of markets will never be better than powerful market players really want it.)

The problems with "applied" business ethics go even further, as far as it is detached from scrutiny of the implicit normativity *of* the market. Such a narrow approach to corporate responsibility is not able to develop a systematic conception of dealing with the inherent necessities of the market economy and the interests behind them, due to the abandonment of reflection in the face of empirically given circumstances (for details see sect. 3.1 in Ulrich 2008). Instead, a blind confidence in the system's harmonic functioning for the common good is still "applied" in principle, apart from situational moral corrections (and the restricted task for business ethics of justifying them). The economistic rhetoric of impersonal and therefore allegedly impartial system necessities remains unquestioned. This problem of anonymous "government of things" or *Sachzwang* is well demonstrated in Galavielle's comment; he makes clear why the primacy of ethically well-justified political governance is essential today. I would like to add: especially on the level of global governance.

A symptomatic consequence of the missing normative criterion of reasonable market "necessities" can be seen in the tendency of applied business ethics to a latent "overadaptation to the neoliberal agenda" (as Losoncz states in his comment) by elevating the business case to the decisive principle for "sound" corporate ethics, instead of the moral principle. It was Richard DeGeorge (1991) who was earlier than others – and most prominently – concerned about this when he asked: "Will Success Spoil Business Ethics?" 15 years later, it is Luk Bouckaert (2006: 208) who again states the "growing use of business ethics as a managerial instrument" and asks for a horizon of business ethics beyond the "given" capitalism. To enrich it with the republican spirit of true liberalism might be a more promising idea than to wait and see until the precarious normative foundations of the whole system will have lost their last legitimating force. The reason for that endeavour is simple in the end: markets cannot "know" for what and for whom they shall be efficient – we have to "tell" them. To do that under the orientating horizon of a politico-philosophically "civilized" conception of business *in* society might be more important than anything else for the practical merit of business ethics in the future.

References

Bouckaert, L. (2006). When More Ethics Creates Less Ethics: Some Further Clarifications on the Ethics Management Paradox. *Interdisciplinary Yearbook of Business Ethics 2006*. pp. 207–210.

Dagger, R. (1997). *Civic Virtues, Citizenship, and Republican Liberalism.* New York, Oxford: Oxford University Press.

DeGeorge, R. T. (1990). *Business Ethics.* 3rd ed. New York: Macmillan.

DeGeorge, R. T. (1991). Will Success Spoil Business Ethics? In Freeman, R.E. (Ed.), *Business Ethics: The State of the Art.* New York, Oxford: Oxford University Press. pp. 42–56.

Kaler, J. (2000). Positioning Business Ethics in Relation to Management and Political Philosophy. *Journal of Business Ethics,* 24. pp. 257–272.

Kaldor, M. (2004). *Global Civil Society: An Answer to War.* Cambridge: Polity Press.

Kant, I. (1991). An Answer to the Question: What is Enlightenment? (orig. 1783). In Kant, *Political Writings*, ed. H. Reiss, transl. by H. B. Nisset, Cambridge Edition of the Works of Immanuel Kant, Cambridge, pp. 54–60.

Kohlberg, L. (1981). *Essays on Moral Development, vol. I: The Philosophy of Moral Development.* San Francisco: Harper & Row.

Maak, Th. & Ulrich, P. (2007). *Integre Unternehmensführung: Ethisches Orientierungswissen für die Wirtschaftspraxis.* Stuttgart: Schäffer-Poeschel.

Marshall, T. H. (1987). *Citizenship and Social Class.* London: Pluto Press (orig. 1950).

Rawls, J. (1993). *Political Liberalism.* Columbia New York: University Press.

Ulrich, P. (2008). *Integrative Economic Ethics: Foundations of a Civilized Market Economy.* Cambridge, New York: Cambridge University Press.

About the Contributors

Laura Albareda

Laura Albareda is researcher at the University Pompeu Fabra (Barcelona) and at the Institute for Social Innovation of ESADE Business School. She has been visiting research fellow at the Centre for the Study of Globalization and Regionalization at Warwick University during 2007–2008. She holds degrees in Philology and Political and Administration Science, and has a Master in Development Studies. She has a Master in International Relations and European Integration from Universidad Autónoma de Barcelona. She is currently a Ph.D. candidate in International Relations.

She is co-author of the book: *Governments and Corporate Social Responsibility* (Gránica, 2005 and Palgrave McMillan, 2008). She has published papers in academic journals including Journal of Business Ethics, Business and Society Review and Corporate Governance: The International Journal of Business in Society. She is co-author of a chapter in the book *Corporate Social Responsibility. Reconciling Aspiration with Application* (edited by A. Kakabadse & M. Morsing, 2005, Palgrave McMillan and European Academy of Business in Society).

Robert Elliott Allinson

Professor Robert Elliott Allinson is the author of over two hundred publications and author or editor of seven books including *Understanding the Chinese Mind*, Oxford University Press, tenth impression, 2000 and *Saving Human Lives*, Springer, 2005. His works have been translated into a number of languages including French, German, Italian, Chinese and Korean. His books have been reviewed in over fifty international publications and academic journals from The Far Eastern Economic Review and Lloyd's List to Philosophy, published by the Royal Society of Philosophy at Cambridge University. Professor Allinson is cited in a number of encyclopedias including the Encyclopedia Britannica and has been invited to be Visiting Professor or Fellow to many universities including Balliol College, St. Antony's College and the Graduate School at Oxford University, Cambridge University, the Graduate School at Yale University, Fudan University in Shanghai, Peking University in Beijing, Waseda University in Tokyo, University of Hawaii, Copenhagen University, University of Canterbury and the Helsinki School of Economics and Business Administration.

He has been Professor for most of his academic career at the Chinese University of Hong Kong, ranked as one of the fifty top universities in the world and is now Professor of Philosophy at Soka University of America.

Gerhold Becker

Gerhold K. Becker, currently Regular Visiting Professor in the Graduate School of Religion and Philosophy at Assumption University of Thailand, Bangkok, taught for nearly 19 years at Hong Kong Baptist University until his retirement in 2004. He was a Chair Professor of Philosophy and Religion and the Founding Director of the university's Centre for Applied Ethics. From 1996–2004 he served as an appointed member on the Hong Kong Council on Human Reproductive Technology of the Government of Hong Kong and from 2001–2004 chaired the Council's Ethics Committee.

Becker is Research Fellow of the Centre for Business Ethics at the Shanghai Academy of Social Sciences, Elected Member of the Association for International Business Ethics, Hong Kong, Founding Member of the Macau Association of Business Ethics as well as of the Ladanyi-Verein zur Förderung der Wirtschaftsethik, Zurich. He is Series Editor of Studies in Applied Ethics (Rodopi, New York/Amsterdam), and an Elected Member of the European Academy of Sciences and the Arts, Salzburg. He holds licentiate degrees in philosophy (Munich) and theology (Frankfurt) and a doctorate in philosophy from the University of Munich, Germany.

Apart from numerous publications in the areas of philosophy and religious studies, Becker has also published on a wide range of issues in applied ethics and in business ethics proper. The following books are representative of his research in applied ethics: *Ethics in Business and*

Society: Chinese and Western Perspectives (1996); *Changing Nature's Course: The Ethical Challenge of Biotechnology* (1996); *The Moral Status of Persons: Perspectives on Bioethics* (2000).

Zsolt Boda

Zsolt Boda was born in 1969 in Budapest, Hungary. He holds an MA in economics and a Ph.D. in political science. He is senior research fellow at the Institute of Political Science, Hungarian Academy of Sciences, and associate professor at the Business Ethics Center, Corvinus University of Budapest. He is also active in the Hungarian green movement and has worked as an expert for environmental NGOs.

He has co-edited and written books in Hungarian on corporate ethics, political theory, environmental politics and policy. He has published several papers in academic journals, and books on international ethics involving the fair trade problematic, trade and environmental issues, and the politics of global environmentalism.

His publications in English include the following:

- Can Governance Structures and Civil–Corporate Partnerships Manage the Global Commons? In Henri-Claude de Bettignies, Francois Lépineux (Eds), *Business, Globalization and the Common Good.* Oxford: Peter Lang, 2008 – forthcoming.
- Hungary. In Naren Prasad (Ed.) (2008) *Social policies and private sector participation in water supply. Beyond regulation* (pp. 178–202). London: Palgrave Macmillan. (With Gábor Scheiring, David Hall and Emanuele Lobina).
- The Ethical Consumerism Movement. *Interdisciplinary Yearbook of Business Ethics,* 1, 2006, pp. 141–153. (With Emese Gulyas).
- Global Environmental Commons and the Need for Ethics. *Society and Economy,* Fall 2003, pp. 213–224.
- Globalization and International Ethics. In Laszlo Zsolnai (Ed.) (2002) *Ethics in the Economy: Handbook of Business Ethics (*pp. 233–258). Oxford: Peter Lang.

Frans de Clerck

Frans de Clerck is co-founder of Triodos Bank Belgium and senior advisor to the Executive Board of Triodos Bank Group. He is chairman of the Institute of Social Banking, board member of Triodos sustainable investment funds and of sustainable companies in Belgium and Luxemburg.

Frans de Clerck was born in 1945 in Belgium. He studied literature and public relations. He has an MBA degree from the Vlerick Leuven Gent Management School and is lecturing on business ethics in finance at this institution and at the Institute for Social Banking. He worked as an advisor to the Council of Europe and the European Commission on social cohesion and corporate social responsibility. He has been working in banking since 1968 and in ethical banking since 1983.

Frank Dixon

Frank Dixon advises businesses, governments and other organizations on sustainability, system change and enhancing financial performance through increased corporate responsibility. For seven years, he was the Managing Director of Research for Innovest Strategic Value Advisors, the largest corporate sustainability research firm in the world. At Innovest, he developed financially-focused models and methods for assessing corporate environmental and social performance and helped institutional investors develop high-performing socially-responsible investing products. His work overseeing the sustainability analysis of the world's 2,000 largest companies made it clear that systemic issues compel all companies to operate unsustainably by making full impact mitigation impossible. To engage business and investors in driving the system changes needed to achieve sustainability, he developed a new sustainability approach focused on system change, called Total Corporate Responsibility (described on www.EuropeSystemChange.com). Before Innovest, he worked in the energy and finance areas. He is advising Wal-Mart and other companies on sustainability. He has an MBA from the Harvard Business School.

Edwin M. Epstein

Edwin M. Epstein is Professor Emeritus at Haas School of Business and International and Area Studies at the University of California, Berkeley. He was Director of the Rotary Center for International Studies in Peace and Conflict Resolution, Associate Dean of International Relations, International and Area Studies at the University of California, Berkeley. From 1994–2001, he was Dean of the School of Economics and Business Administration and Earl W. Smith Professor at Saint Mary's College of California. He is a member of the California, Pennsylvania, and Supreme Court of the United States Bars.

Professor Epstein is noted internationally for his efforts to infuse ethical inquiry and socio/political analysis into management education. His 40 years of teaching and scholarship have earned him significant professional recognition. His book The Corporation in American Politics won the Howard Chase Book Award from the Social Issues in Management Division of the Academy of Management for making "a significant and lasting contribution to the study of business and society." He is a former Chair of the Academy's Social Issues in Management Division and recipient of the Division's Summer Marcus Distinguished Service Award, and a past Fellow of the Woodrow Wilson International Center for Scholars in Washington, D.C.

Laszlo Fekete

Laszlo Fekete is associate professor of the Business Ethics Center at the Corvinus University of Budapest. He was born in 1951, in Budapest, Hungary. He studied history, economic history, and sociology at Eötvös Loránd University of Budapest, the Friedrich-Schiller-Universität zu Jena (Germany), and State University of New York at Binghamton (USA). He has MA in economic history and sociology, and a Ph.D. in philosophy. His research interests include the philosophical and ethical problems of business transactions, digital culture and information society. His recent essays – *Man, Machines, and Communications*;

Rights, Rules, and Regulations in Cyberspace; *The Networks of Philosophy*; *Ethics of Economic Transactions in the Global Network Society* – have been published in different philosophical, economic and sociological reviews at home and abroad. He is the co-author and editor of *Contemporary Ethics* published in 2004. He is one of the contributors to the book on corporate social responsibility – *Corporate Social Responsibility across Europe* – which has been published by Springer Verlag.

Jean-Pierre Galavielle

Jean-Pierre Galavielle has been Associate Professor and accreditated as Research Superviser at the University Paris-I Panthéon-Sorbonne. His research interests include the economics of public finance, taxation and employment and economics and industrial finance. In 1982–1997 he was a member and Vice-President of the National University Council. In 1988–1993 he served as Vice-President of University Paris-1, in charge of the policy of scientific research. In 1994–2000 he was Director of University Center Pierre Mendès-France (Tolbiac), University Paris-I. His recent publications include

- Arnal J. & Galavielle, J.-P. (2006). Confiance et marché, entre norme éthiques et puissance publique, *European Journal of Economic and Social Systems, "Ethique, Economie et Société: Une affaire de politique?"* Vol. 19, No. 1/2006.
- Galavielle, J.-P. (2004). Business Ethics Is a Matter of Good Conduct and of Good Conscience?" *Journal of Business Ethics*, Vol. 53, August 2004. Kluwer.
- Galavielle, J.-P. (2004). Esiste una teoria dei mercati finanziari? *Global and Local Economic Review*, Vol. No. VII/2004. Pescara: Caripe-Edizioni Tracce.

Eleanor O'Higgins

Eleanor O'Higgins (B.A., M.Sc., M.B.A., Ph.D.) is on the faculty of the Business Schools at UCD and a Visiting Fellow at the London School of Economics and Political Science. She specialises in teaching, research and publications in business ethics, corporate social responsibility, corporate governance and strategic management. She is the author of numerous papers in academic and professional journals, newspaper articles, book chapters and case studies. Her case study and research work on the airline industry and on Ryanair have won numerous international awards.

She serves on various boards and bodies. Among these, she is a member of the Press Council of Ireland, a director of Transparency International Ireland, and of The Marine Institute; and a member of the Board of Management of the Centre for Corporate Governance at University College Dublin. She is a member of the Business Ethics and of the Public Management & Governance interfaculty groups of the Community of European Management Schools (CEMS) and has held various leadership positions in the US Academy of Management. She is on the editorial boards of a number of international management journals. Eleanor carries out various teaching and speaking assignments internationally.

She has extensive experience in business through consulting work, her previous managerial career, and as a company director. Previously, she worked as a research psychologist at Harvard University Medical School, as a clinical psychologist in the Irish health services, and as staff development manager in RTE, the Irish national broadcasting organisation. She has served on the board of directors of IDA Ireland and of the Well Woman Group of health clinics.

Benedek Javor

Benedek Javor was born in 1972 in Budapest, Hungary. He received MSc in biology from ELTE University, Hungary in 1997 and Ph.D. in Ecology and Zoology in 2006. He also studied cultural anthropology and human ecology from 1993 to 1997. He is assistant professor of environmental sciences at the department of environmental law, Pazmany Peter Catholic University, Budapest. Founding member of the environmental NGO 'Vedegylet – Protect the Future Society' in 2000; secretary of the society's 'Representation of Future Generations' (REFUGE) program 2000–2002, spokesperson 2003–2008. Benedek Javor was the leader of the organization's lobby-work for the creation of a parliamentary guardian (ombudsman) for future generations till the official creation of the institution in 2007. He serves as the program manager of the "European representation of future generations" project, run since 2005. Main research areas: environmental ethics, religion and ecology, rights of future generations, public participation, urban environment, forest management and nature conservation. His publications include:

- Javor, B., Varady, T., and Toma, G. (2006). *Before You Lock Yourself There – A Guide for the Protection of Urban Trees and Green Areas.* Budapest: Vedegylet (in Hungarian).
- Exner, T. & Javor, B. (2003). *Forest Survey Report 2003.* Budapest: WWF (in Hungarian).
- Institutional Protection of Succeeding Generations – Ombudsman for Future Generations in Hungary. In J. Tremmel (2006), *Generational Justice Handbook.* Cheltenham, UK: Edward Elgar Publishing.
- Representing Future Generations – A Question of Democracy for Europe, *Mechanism of Economic Regulation*, 1/2008.
- Smaller State, A Reform Failed? – Assessment Report of the Transformation of the Environmental Authorities in Hungary. *Társadalomkutatás*, Vol. 25, No. 4, November 2007, (in Hungarian).
- A Proposal on the Representation of Future Generations in Hungary. In B. Javor & J. Racz (2006), *Do We Owe Them a Future? – The*

Opportunities of a Representation for Future Generations in Europe. Budapest: Vedegylet.

- Hungary: Forest Degradation with State Assistance. *EU Forest Watch*, 93, February 2005.

Alpar Losoncz

Alpar Losoncz was born in 1958. in Temerin, Serbia. He has studied law, economics, and philosophy. He gained his Ph.D. in Social Sciences and Economics from the University of Novi Sad. Since 2005 he is university professor. Since 1996 he serves as the visiting professor at the Faculty of Arts in Novi Sad, and since 1991 he serves as visiting professor at the Faculty of Arts in Szeged (Hungary), too. He serves as lecturer at the Institute for the Engineering and the Environmental Protection of University of Novi Sad and he is the member of Scientific Council of the same University.

Dr. Losoncz has more than 100 publications on interrelated topics of economics, business ethics, ecology and philosophy. His interest is the ecology in interdisciplinary frameworks and the analysis of the interactions between the socio-ecological dynamics and business ethics. Recently he has published a book on the coevolution between economy and the environment treating the socio-economic aspects of ecological structures. His most important publications include:

- *Suffitientia ecologica*. 2005. Novi Sad.
- *Sovereignty, Power, and Crisis.* 2006.
- Spirituality and Motivation. In L. Zsolnai (Ed.), *Spirituality and Ethics in Management.* 2004 (pp. 75–87). Dordrecht, Boston, London: Kluwer Academic Publishers, Issues in Business Ethics.
- Confronting a Leadership Vacuum. In A. Habisch (Ed.), *Corporate Social Responsibility, Across Europe, Discovering National Perspectives of Corporate Citizenship*. 2004 (pp. 233–244). Heidelberg: Springer Verlag.

Josep M. Lozano

Josep M. Lozano was awarded a Ph.D. in Philosophy by the University of Barcelona and a degree in Theology by the Theology Faculty of Catalonia. He also holds a degree in Executive Management from ESADE Business School. He is currently Full Professor at the Department of Social Sciences at ESADE and Senior Researcher in CSR at the Institute for Social Innovation (ESADE). His academic and professional activity focuses on the fields of Applied Ethics; Corporate Social Responsibility; and values, leadership and spirituality.

Josep Lozano's major books include *Governments and Corporate Social Responsibility* (Palgrave); *Tras la RSE. La RSE en España vista por sus actors* (Granica); *Guía práctica de Derechos Humanos para empresas* (UAB); and *Ethics and Organizations. Understanding Business Ethics as a Learning Process* (Kluwer), which received the Joan Sardà Dexeus Prize in 1998 as the best book on corporate economics published in Catalonia (pub. Proa). The work was also published in Spanish (pub. Trotta). Lozano has also collaborated in a further 39 books. He has been Director of the Observatory on Ethical, Ecological and Social Funds in Spain, which has so far published seven annual reports.

He is co-founder of *Ética, Economía y Dirección* (the Spanish branch of the European Business Ethics Network); member of the editorial board of the journal *Ethical Perspectives* and *Society and Business Review*, and a member of the boards of trustees and advisory councils of various foundations and associations. Josep M. Lozano was the ESADE representative in the consortium of European universities that created the European Academy for Business in Society (EABiS). He was member of the Catalan Government's Commission on Values, and of the Spanish Ministry of Employment and Social Affairs' Commission of Experts on Corporate and Social Responsibility. Prof. Lozano was the highly commended runner-up for the Faculty Pioneer Award (2003), held by EABiS and The Aspen Institute.

Atle Midttun

Atle Midttun is Professor at the Norwegian School of Management and Director of The Centre for Corporate Citizenship.

He got his Masters Degree form the University of Oslo and his Ph.D. from the University of Uppsala. He was Visiting Scholar at the Max Planck Institut für Gesellschafts-forscung, Köln (1988), Guest Researcher at the University of Aalborg (1995), Visiting Professor, University of Michigan, Business School/School of Natural Reosurces (2000–2001), Visiting Professor at Université Paris Sud (2002) and Visiting Scholar at the University of California, Berkeley, Haas School of Business (2003–2004).

His special fields are economic organisation, regulation theory, organisation theory, innovation theory corporate social responsibility, political and institutional economics, and methodology, energy- and environmental policy/economics, corporate social responsibility, industrial policy, innovation policy, business strategies, public sector management.

Laura Nash

Laura L. Nash, has been an educator, consultant, and author on business ethics, leadership and values for over twenty-five years. She recently left the faculty of Harvard Business School, where she taught in the entrepreneurship unit, to pursue other career interests. She continues to consult on values and leadership, and serves on the senior academic advisory committee of the Business Roundtable's Institute for Ethics' advisory board, and is also a senior fellow at Bentley College Center for Business Ethics' executive fellow program.

Her most recent book, with Prof. Howard Stevenson, is *Just Enough. Tools for Creating Success in Work and Life*, (John Wiley & Sons Inc. 2004). Prior books include: *Good Intentions Aside* (Harvard Business School Press), *Believers in Business* (Thomas Nelson Publishers), *Church on Sunday, Work on Monday*, with Scotty McLennan (Josey-Bass Publishers), *Policies and Persons: A Casebook in Business Ethics* (with Kenneth Goodpaster et al.). She has worked with many for-profit and not-for-profit organizations from The World Economic Forum to PepsiCo to Young Presidents Organization. Past president of the Society for Business Ethics, she was program director for The Conference Board's first conferences on business ethics, and she has written several cover articles for *Across the Board*, including "The Nanny Corporation" and "The Evangelical CEO".

Laura Nash received a Ph.D. and M.A. in classical philology from Harvard University and graduated from Connecticut College in 1970. She lives in Cambridge, Massachusetts with her husband and two daughters.

Francesco Perrini

Francesco Perrini is Full Professor of Management and CSR at the Institute of Strategy, Department of Management, Bocconi University, Milan, Italy. He is also SIF Chair of Social Entrepreneurship and Senior Professor of Corporate Finance at the Corporate and Real Estate Finance Department, SDA Bocconi School of Management. Prof. Perrini is currently Director of Bocconi CSR Unit, Department of Management, Bocconi University, and Coordinator of CSR Activities Group at SDA Bocconi.

His research areas:

- *Management of Corporate Development Processes:* from strategy implementation (acquisitions and strategic alliances) to financial strategies and valuation.
- *Small-Medium Enterprises.*
- *Social Issues in Management:* corporate governance, corporate social responsibility (CSR), sustainability, social entrepreneurship, social innovation, socially responsible investing (SRI).

Francesco Perrini's publications include the following:

- Perrini, F., Castaldo, S., Misani, N., & Tencati, A. (2008). "The Missing Link between Corporate Social Responsibility and Consumer Trust: The Case of Fair Trade Products", *Journal of Business Ethics*, forthcoming.
- Perrini, F., Russo, A., & Tencati, A. (2007). "CSR Strategies of SMEs and Large Firms. Evidence from Italy", *Journal of Business Ethics*, 74(3), pp. 285–300.
- Perrini, F. (2006). "The Practitioner's Perspective on Non-Financial Reporting", *California Management Review*, 48(2), pp. 73–103.
- Perrini F. (2006). "SMEs and CSR Theory: Evidence and Implications from an Italian Perspective", *Journal of Business Ethics*, 67(3), pp. 305–316.
- Perrini, F. (Ed.) (2006). *The New Social Entrepreneurship. What Awaits Social Entrepreneurial Ventures?*, "Preface: Entrepreneurship and Societal Wealth Generation", by Ian MacMillan (Wharton, University of Pennsylvania), Cheltenham, UK: Edward Elgar Publishing Ltd., pp. 368.

Yvon Pesqueux

Yvon Pesqueux is Professor at CNAM (Conservatoire National des Arts et Métiers, Paris), where he is Head of the Chair "Développement des Systèmes d'Organisation".

Yvon Pesqueux is born in 1951. Ha has a Ph.D. in Economics from University of Paris 1 Panthéon-Sorbonne (1975). He is editor of Society and Business Review (Emerald Publishing: www.emeraldinsight.com/sbr.htm) and serves as member of several editorial boards. His books include the following:

- *Le gouvernement de l'entreprise comme idéologie* (2000). Paris: Ellipses;

- *Organisations: modèles et représentations* (2002). Paris: P.U.F.;
- *Ethique des Affaires* (2002). Paris: Editions d'Organisation;
- *L'entreprise multiculturelle* (2004). Paris: L'Harmattan;
- *La dérive organisationnelle* (2004). Paris: L'Harmattan, 2004 (with Bruno Triboulois);
- *L'organisation en réseau: mythe ou réalité.* (2004). Paris: P.U.F., collection "la politique éclatée". (In collaboration with Michel Ferrary);
- *Stakeholder Theory: A European Perspective* (2005). Basingstoke, UK: Palgrave-Macmillan. (In collaboration with Maria Bonnafous-Boucher);
- *Management de la connaissance* (2006). Paris: Economica. (In collaboration with Michel Ferrary).

James Robertson

James Robertson's early career was in government in London: with Prime Minister Harold Macmillan on his "Wind of Change" tour of Africa in 1960, then in the Cabinet Office. After that, director of inter-bank research for the big British banks.

In the 1970s he became an independent writer, speaker, and consultant on economic and social change. In the mid-1980s he co-founded The Other Economic Summit (TOES) and the New Economics Foundation, London. His books include the following (details at http://www.jamesrobertson.com/books.htm.):

- *Reform of British Central Government.* 1971. London: Chatto & Windus/Charles Knight.
- *Profit or People? The New Social Role of Money.* 1974. London: Marion Boyars.
- *Power, Money and Sex: Towards a New Social Balance.* 1976. London: Marion Boyars.

- *The Sane Alternative.* 1978, 1983. London, Shropshire, UK: James Robertson.
- *Future Work: Jobs, Self-employment and Leisure After the Industrial Age.* 1985. Aldershot, UK: Temple Smith/Gower.
- *Future Wealth: A New Economics for the 21st Century.* 1990. London: Cassell.
- *Beyond The Dependency Culture: People, Power and Responsibility.* 1998. London: Adamantine Press.
- *Transforming Economic Life: A Millennial Challenge.* 1998. Schumacher Briefing No. 1, Green Books, UK.
- *The New Economics of Sustainable Development: A briefing for policy-makers.* 1999. Written for the European Commission). Published by: Kogan Page, London. Editions Apogee, Paris (as Changer d'Economie: ou la Nouvelle Economie du Developpement Durable), and Office for Official Publications of the European Communities, Luxembourg.
- *Creating New Money: A Monetary Reform for the Information Age* (with Prof Joseph Huber). 2000. London: New Economics Foundation.
- *Monetary Reform – Making it Happen.* 2004. London: International Simultaneous Policy Organisation. (With John Bunzl).

J. M. Sampath

J. M. Sampath, is an intensive reflective thinker and comes with a vast experience of having worked with several organizations globally, including some Fortune 100 companies. His core skills are leadership development and coaching. He works with organizations in nurturing a culture of excellence and building leadership. He heads his own consulting company with offices in India, Malaysia, Singapore and the US. He has authored two books and published several papers. He has also developed the Vision Value Instrument and Value Profile Instrument, both of which have been extensively used in individual, team, and organizational diagnosis and coaching over the last decade. Dr. Sampath has addressed several conferences, in India and abroad, and his presentations on "Learning for Nurturing Excellence" and "Vision Value Alignment" have been widely appreciated in different countries. Dr. Sampath has been one of the founder promoters of Kshema Technologies, a successful technology startup.

Antonio Tencati

Antonio Tencati is Assistant Professor of Management and Corporate Social Responsibility at the Department of Management – CSR Unit of Bocconi University in Milan.

He was born in 1968. He has a Degree in Business Administration at Bocconi University. Since 1993 he has been a researcher at SPACE (the European Research Centre of Bocconi University on Risk, Security, Occupational Health and Safety, Environment and Crisis Management). Since 2000 he has been a member of the Business Ethics Faculty Group of the CEMS (Community of European Management Schools). Since 2005 he has been an Assistant Professor at Bocconi University.

Antonio Tencati's research areas include Business Management, Management of Sustainability and Corporate Social Responsibility, Environmental Management, Innovation and Operations Management.

His publications include the following:

- Tencati, A. & Zsolnai, L. (2008). "The Collaborative Enterprise", *Journal of Business Ethics*, forthcoming.
- Russo, A. & Tencati, A. (2008). "Formal vs. Informal CSR Strategies: Evidence from Italian Micro, Small, Medium-sized, and Large Firms", *Journal of Business Ethics*, forthcoming.
- Pivato, S., Misani, N., & Tencati, A. (2008). "The Impact of Corporate Social Responsibility on Consumer Trust: The Case of Organic Food", *Business Ethics: A European Review*, 17(1), pp. 3–12.

- Lenssen, G., Perrini, F., Tencati, A., & Lacy, P. (Eds) (2007). "Corporate Responsibility and Strategic Management", *Corporate Governance: The International Journal of Business in Society*, 7(4).
- Perrini F., Pogutz, S., & Tencati, A. (2006). *Developing Corporate Social Responsibility. A European Perspective.* Cheltenham, UK: Edward Elgar Publishing.

Paula Tiihonen

Paula Tiihonen was born in the middle of Finland, studied there administrative sciences and after graduation moved to Helsinki. During 30 years she has worked and continued studies. She doctorated in 1986 in administrative sciences, when working as a Counsellor of Legislation in the Ministry of Justice.

Last 15 years she has been working in the Finnish Parliament, Counselor for the Economic Affairs Committee and the Committee for Labor Affairs and then since 1989 for the Committee for the Future.

She has written several books on public administration. Her latest publications include:

- Mannermaa, M., Dator, J. & Tiihonen, P. (Eds) (2006). *Democracy and Futures.* The Finnish Parliament.
- Kuusi, O., Smith, H., & Tiihonen, P. (Eds) (2007). *Russia 2017: Three Scenarios.* The Finnish Parliament.

Peter Ulrich

Peter Ulrich is Full Professor of Economic and Business Ethics and Director of the Institute for Business Ethics at the University of St Gallen, Switzerland. He was born 1948 in Berne, Switzerland, studied business administration, economics and social sciences at the University of Fribourg (Switzerland) and did his doctorate on "The corporation as a quasi-public institution" at the University of Basel (1976). After having worked four years as a management consultant in Zurich, he completed his habilitation thesis on "The transformation of economic reason", which was accepted by the University of Witten-Herdecke (1986). He got his first appointment as a full professor for Business Administration at the University of Wuppertal, Germany (1984–87). 1987 he received the call to St. Gallen, where he founded the Institute for Business Ethics (1989) and was dean of the Humanities and Social Sciences Department (2005–07). He has developed the approach of "integrative economic ethics" to one of the most-discussed conceptions in the German debate. Among his 27 books are:

- *Transformation der ökonomischen Vernunft.* 1986. 3rd ed. 1993. Bern, Stuttgart, Vienna: Haupt Publ.
- *Ethik und Erfolg. Unternehmensethische Denkmuster von Führungskräften* (with U. Thielemann). 1992. Bern, Stuttgart, Vienna: Haupt Publ.
- *Facing Public Interest. The Ethical Challenge to Business Policy and Corporate Communications* (co-ed. with Ch. Sarasin). 1995. Boston, Dordrecht, London: Kluwer Academic Publ.
- *Integrative Wirtschaftsethik. Grundlagen einer lebensdienlichen Ökonomie.* 1997. 4th rev. ed. 2008. Bern, Stuttgart, Vienna: Haupt Publ.
- *Der entzauberte Markt. Eine wirtschaftsethische Orientierung.* 2002. Herder Publ., Freiburg (Germany); rev. paperback edition: *Zivilisierte Marktwirtschaft,* 2nd and 3th ed. 2005.
- *Brennpunkt Bankenethik. Der Finanzplatz Schweiz in wirtschaftsethischer Perspektive* (with U. Thielemann), 2003.
- *John Stuart Mill: Der vergessene politische Ökonom und Philosoph* (co-ed. with M.S. Assländer). 2006. Bern, Stuttgart, Vienna: Haupt Publ.
- *Kritika Ekonomisma.* 2004. Moskva: Vusovskaja Kniga Publ.
- *Integrative Economic Ethics. Foundations of a Civilized Market Economy.* 2008. Cambridge, New York: Cambridge University Press.
- *Ética económica integrativa. Fundamentos de una economía al servicio de la vida.* 2008 (in print). Quito, Ecuador: Abya-Yala Publisher.

Imre Ungvari-Zrinyi

Imre Ungvari-Zrinyi is associate professor in General and Applied Ethics and Axiology at the "Babes-Bolyai" University Cluj and teaching Negotiation and Media Ethics at the "Sapientia" University, Targu Mures, Romania.

Imre Ungvari-Zrinyi was born in 1960, in Targu Mures, Romania. He graduated in Philosophy at the "Babes-Bolyai" University Cluj in 1985 and got his Ph.D. in Philosophy of Values at the same university in 1999. He is member of the Diotima Philosophical Association in Cluj and the Hungarian Philosophical Association in Budapest.

Imre Ungvari-Zrinyi is editorial board member of the de Philosophical Review "Kellek" in Cluj. His books include the following:

- *Self-positioning and Consciousness of Value. The Philosophy of Karoly Bohm.* 2002. Cluj: Pro-Philosophia. (In Hungarian)
- *Dialog. Interpretation. Interaction. Toward a Communicative Approch to Culture.* 2005. Targu Mures: Mentor (In Hungarian)
- *Introduction in Ethics.* 2006. Bucharest: Ed. Didactica si Pedagogica (In Hungarian)
- *Basic Concepts of Applied Ethics. Bioethics, Business Ethics, Ethics of Public Service, Media Ethics.* 2007. University Workshop, Bolyai Society, Cluj. (In Hungarian)

Jakob von Uexkull

Born in Uppsala, Sweden, Jakob von Uexkull is the son of the author and journalist Gösta von Uexkull and grandson of the biologist Jakob von Uexkull. He graduated with an M.A. (Honors) in Politics, Philosophy and Economics from Christ Church, Oxford. He holds both Swedish and German nationality.

Jakob von Uexkull has been a journalist, writer, lecturer, professional philatelist and Member of the European Parliament (1987–89). As a Member of the European Parliament, he served on the Political Affairs Committee and the Science and Technology Committee. He was also a member of the Delegation for Relations with the Supreme

Soviet of the USSR and the Baltic Intergroup. Jakob von Uexkull is a co-founder of The Other Economic Summit (TOES) (1984) and founder of The Estonian Renaissance Award (1993). He served on the Board of Greenpeace, Germany, and the New Economics Foundation (London). He was also a member of the UNESCO Commission on Human Duties and Responsibilities.

Jakob has been honored by Time Magazine as a European Hero (2005). In 2006, he received the Binding-Prize (Liechtenstein) for the protection of nature and the environment, and, in 2008, the Erich-Fromm-Prize in Stuttgart, Germany, for his future-orientated commitment dedicated for a life in human dignity for our children and grand-children in a globalised world.

Stephen B. Young

Stephen B. Young is the Global Executive Director of the Caux Round Table, an international network of experienced business leaders who advocate a principled approach to global capitalism. Young has recently published *Moral Capitalism*, a well-received book written as a guide to use of the Caux Round Table ethical and socially responsible Principles for Business.

Young was born in 1945 in Washington, D.C. He was educated at the International School Bangkok, Harvard College and Harvard Law School. Young volunteered for service during the Vietnam War. After a year of language training, he worked in village development for the CORDS Advisory program in South Vietnam from 1968 to 1971. Young has subsequently written several articles on Vietnamese nationalism. With the late Prof. Nguyen Ngoc Huy, Young published a book study of human rights in traditional China and Vietnam. In 1975 Young took the initiative to begin the resettlement program of refugees from Indochina after the Vietnam War was lost.

He came to Minnesota in 1981 to be the third dean of the Hamline University School of Law. Previously, he had been an Assistant Dean at Harvard Law School. Young has also taught at the University of Minnesota Law School, Vietnamese history for the College of Liberal Arts, University of Minnesota and Public Office as a Public Trust for

Minnesota State University – Mankato. He has published articles on Chinese jurisprudence, the culture and politics of Vietnam and Thailand, legal education, law firm management, Native American law, the history of negligence, and the law of war. His most recent article on legal philosophy discusses the morality of American law.

Laszlo Zsolnai

Laszlo Zsolnai is professor and director of the Business Ethics Center at the Corvinus University of Budapest. He is chairman of the Business Ethics Interfaculty Group of the Community of European Management Schools (CEMS).

Laszlo Zsolnai was born in 1958. He has a master's in finance and a doctorate in sociology from the Budapest University of Economic Sciences. He received his Ph.D. and DSc degrees in economics from the Hungarian Academy of Sciences. His books include the following:

- *The European Difference.* 1998. Boston, Dordrecht, London: Kluwer Academic Publishers.
- *Ethics and the Future of Capitalism.* 2002. New Jersey, London: Transactions Publisher. (With Wojciech Gasparski).
- *Ethics in the Economy: Handbook of Business Ethics.* 2002. Oxford, Bern: Peter Lang Academic Publisher.
- *Spirituality, Ethics and Management.* 2004. Boston, Dordrecht, London: Kluwer Academic Publishers.
- *Business Within Limits: Deep Ecology and Buddhist Economics.* 2005. Oxford, Bern: Peter Lang Academic Publisher. (with Knut J. Ims).
- *Spirituality as a Public Good.* 2007. Antwerp, Apeldoorn: Garant. (With Luk Bouckaert)
- *Frugality: Rebalancing Material and Spiritual Values in Economic Life.* 2008. Oxford: Peter Lang Academic Publishers. (With Luk Bouckaert and Hendrik Opdebeeck)
- *Europe–Asia Dialogue on Business Spirituality.* 2008. Antwerp, Apeldoom: Garant.

GPSR Compliance

The European Union's (EU) General Product Safety Regulation (GPSR) is a set of rules that requires consumer products to be safe and our obligations to ensure this.

If you have any concerns about our products, you can contact us on ProductSafety@springernature.com

In case Publisher is established outside the EU, the EU authorized representative is:

Springer Nature Customer Service Center GmbH
Europaplatz 3
69115 Heidelberg, Germany

Batch number: 10823775

Printed by Printforce, the Netherlands